```
            WITHDRAWN                    COPY 89
622.19   Sullivan, George, 1927-
  S         The modern treasure finder's manual /
         George Sullivan. 1st ed. Radnor, Pa. :
         Chilton Book Co., [1974]
            185p.  Bibl. 8.95
                              MELVINDALE PUBLIC LIBRARY
                                  18650 ALLEN ROAD
     L                         MELVINDALE, MICHIGAN 48122
    2/15                             DU 1-8677

           1.Treasure-trove.               31225
                                                  J175
```

THE
MODERN
TREASURE FINDER'S
MANUAL

# THE MODERN TREASURE FINDER'S MANUAL

## GEORGE SULLIVAN

CHILTON BOOK COMPANY
*Radnor, Pennsylvania*

Copyright© 1975 by George Sullivan

FIRST EDITION   *All Rights Reserved*

Published in Radnor, Pa., by Chilton Book Company
and simultaneously in Don Mills, Ontario, Canada,
by Thomas Nelson & Sons, Ltd.

Designed by Donald E. Cooke

Manufactured in the United States of America

**Library of Congress Cataloging in Publication Data**

Sullivan, George, 1927-
   The modern treasure finder's manual.

  Bibliography: p.
  Includes index.
  1. Treasure-trove.   I. Title.
G525.S89   1974            622.1'9         74-23975
ISBN 0-8019-6041-X

# CONTENTS

      Introduction . . . *1*
1. The Equipment You Need . . . *11*
2. How to Use It . . . *35*
3. Where to Search . . . *45*
4. All About Coins and How to Find Them . . . *64*
5. What Else to Look For . . . *93*
6. All About Gold and Prospecting . . . *110*
7. Underwater Search . . . *144*
      Appendix . . . *165*
          Books for Treasure Hunters . . . *165*
          Treasure-Hunting Magazines . . . *166*
          Treasure-Hunting Clubs . . . *166*
          States Offering Information on Prospecting and
              Mineral Collecting . . . *169*
          Localities in the United States where Gem
              Materials are Found . . . *177*
      Index . . . *185*

# ACKNOWLEDGMENTS

Many people were helpful in providing source material and photographs used in this book, and it would be impossible to name them all. But special thanks are owed to Charles Garrett, president, Garrett Electronics, who checked portions of the manuscript for technical accuracy; and to A. J. Dumais, president, New England Treasure Finders Association, Dick Boisseau, Aime LaMontagne, Bill Sullivan and Tim Sullivan, who served as photographic models.

# THE MODERN TREASURE FINDER'S MANUAL

# INTRODUCTION

Not long ago, a Rochester, New York, man, kneeling in one of the city's grassy parks, sunk a metal probe that he had fashioned out of a screwdriver into the soft earth. At about five inches, he felt its tip strike something small and metallic.

Taking a thick-bladed hunting knife from his belt, he carved out a cone-shaped plug of soil surrounding the object. When he flipped back the soil, a coin slightly larger than a quarter and blackened by age fell free.

A penny, the coin bore a Liberty head and the date 1841. It was, numismatically speaking, a "large cent." It was worth about $15.

Searching for military artifacts in New York State. *(George Sullivan)*

But Harry Tiernan, the man who unearthed the coin, would never think of selling it. It would go into his collection of more than 7,000 other coins, mostly pennies bearing recent dates, but several more valuable than this one, all of which he has found in city parks, playgrounds, and on Lake Ontario beaches. Besides the coins, he has well over a hundred rings to show for his efforts, one with a diamond that has been appraised at $250; bracelets and other jewelry; medals and transit tokens; forks, spoons, and knives; keys—so many he's stopped counting; metal buttons, half a dozen pocketknives, and a .32-caliber pistol.

Tiernan gave up fishing and golfing on weekends two years ago to join the growing number of amateur treasure hunters, men and women who prowl beaches, parks, schoolyards, and abandoned buildings in search of coins, jewelry, historical relics—anything of value.

The principal tool they use is an electronic metal detector, a device that looks like a walking cane, but with a dish-sized search loop mounted near the tip and a small control box at the handle. The detector generates an electro-magnetic field that penetrates the earth, and when the field strikes a metal object a speaker "sounds off." Then you probe and dig.

The detector is so sensitive that it can perceive a dime at six inches, a horseshoe at a foot and a half, and a metal chest at three feet or even deeper. These instruments, also called metal locators, have done for treasure hunting what the gasoline engine did for surface transportation.

Detectors are now standard equipment in gold prospecting, for they are capable of searching out nuggets and deposits of black magnetic sand wherein gold flakes and minute gold particles are invariably found. Other models are especially meant for underwater use, their control systems encased in waterproof, pressure-proof housings.

Detectors are a cinch to operate. If you have the mechanical ability to tune in a color television set, you can become skilled as a detector user. And they are reasonably priced, ranging from $50 or so, up to $250 or $300, the amount you pay for a deluxe model.

Treasure hunting with electronic detectors began to surge in popularity in the early 1960s, starting in the West and spreading across the country. In recent years treasure hunters have begun to appear up and down the Atlantic Seaboard.

Hardly a week goes by that someone doesn't find something of important value. Consider:

A pair of Michigan women invested $149 in a metal detector because they had been hearing all their lives that grandpa had buried a strongbox containing cash and valuables "somewhere in the garden." It took them

Coinshooting. *(Daytona Beach Resort Area)*

three hours of searching to find it, buried two feet down amidst some grapevines. It contained $700 in old coins and bills.

A Wilton, Connecticut, antiques dealer, digging in a swamp only two miles from his home, unearthed a 20-pound fragment of what the *New York Times* described as "the most famous piece of sculpture in early American history," a large equestrian statue of George III, the last king of the American colonies. Cast in London, the statue was erected in New York's Bowling Green Park in 1770. Angry citizens toppled the statue and hacked it to pieces in 1776, and then shipped the scrap to Litchfield, Connecticut, for melting into bullets. But some pieces never got there and began turning up in Wilton during the 1800s. They are highly prized by museums and historical societies.

In terms of dollar value, one of the biggest recorded finds belongs to an Arkansas woman whose detector chanced upon a metal-capped fruit jar buried three-feet deep in her front yard. It contained silver dollars,

A relic hunt in Massachusetts. *(George Sullivan)*

several Series E Savings Bonds, and a deed that once belonged to her father and that gave her title to a section of wheatland in Angora, Nebraska. The find was worth $87,000.

For sheer volume, there's the case of a staff reporter for the Parkersburg (West Virginia) *News*. In three years of treasure hunting with a $269 detector, he recovered more than 26,000 coins, rings, bracelets, pieces of silverware, and historical artifacts ranging from Model T hubcaps to those conical Civil War rifle bullets known as minie balls. "I found enough rare coins in my first three months of collecting to reimburse me for what I paid for the detector," he says.

When it comes to gold seeking, there's the story of a blond girl and her husband who drove up each weekend to Maxwell Creek in Mariposa County, California, about 150 miles due east of San Francisco. Working in water that was often hip deep, the man would probe the streambed for black-sand deposits, which his wife would pan. They would begin early in the morning and work steadily until nightfall. They did it month after month through one summer, fall, and winter, and into the spring. Their labor earned them a row of small glass vials, each one about the size of a roll of dimes, and filled with gold particles, about $20,000 worth.

A San Francisco treasure hunter keeps on the lookout for downtown lots that have been cleared for redevelopment, for they yield relics from buildings that were destroyed by the San Francisco earthquake in 1906. He has found scores of Chinese coins, military buttons, porcelain beer-bottle stoppers, clay pipe bowls, nineteenth-century bottles, and a well-rusted pocket watch, its hands fixed at 5:12, the exact time in the morning when the earthquake struck.

Despite the spectacular finds you hear about, treasure hunting is not a get-rich-quick type of enterprise. Being successful requires patience and persistence. Don't get any ideas you're going to be able to quit your job.

More than a few treasure seekers will tell you that the income they derive from their activity in the form of the coins and the occasional valuable relics they find is not the greatest benefit. The excitement is. Even veterans claim the pulse quickens whenever the detector gives out a loud "beep." Maybe it's a gold coin or a piece of valuable jewelry, or maybe it's just junk, junk often being those metal pulltabs from soft-drink or beer cans.

You never know what you're going to find. A young man from Boston was coinshooting (a word treasure hunters use to describe what a coin hunter does) at a Vermont ski resort one sunny afternoon in the early spring. He passed the search loop over a melting patch of snow and got a very loud signal. When he kicked the snow away, his foot struck an untouched six-pack of Milwaukee-brewed beer. Every treasure hunter can give you a long list of bizarre objects that he's uncovered.

    This book is to help get you started. It explains how to select a detector that's best suited for your needs, getting the most quality for the least amount of money. It tells you how to obtain peak performance from your machine, whether you're planning to coinshoot in your backyard, probe ownerless buildings or ghost towns, seek deposits of placer gold in the Colorado Rockies, or dive for coins and relics from wrecked Spanish galleons off the Florida coast. Whatever—good hunting.

# Chapter 1
# THE EQUIPMENT YOU NEED

Late in the 1920s at an air station near Sunnyvale, California, Dr. Gerhardt Fisher, under contract to the Navy, tried to perfect a radio direction finder which employed a rotatable loop antenna. The unit operated perfectly, except when the antenna happened to have nearby Redwood City within its range.

Dr. Fisher worked for days trying to solve the problem. Then he discovered that the city's metal water tank, mounted atop a tall tower, was the source of the interference.

After Dr. Fisher had completed his research work for the Navy, he turned his attention back to the electronic phenomenon caused by the Redwood City water tower. Out of the development work that followed, sprang a practical metal detector, one that bears a direct relationship to most of those in use today.

The first models were big and cumbersome and you had to have a doctorate in electronics engineering to be able to make one work. They were a far cry from the models in use today, which usually weigh no more than two or three pounds and are a cinch to operate.

Virtually all makes and models have the same general appearance. Two sets of components are linked by a telescoping metal rod. At one end there's a small handle and a control case about the size of a shoebox. It contains the detector's electronic circuitry and the batteries that power the machine. Mounted on the control box's upper face are control knobs and usually a meter upon which the intensity of the signals received are measured. The control case also contains a speaker which emits audio signals.

At the other end of the unit, there's a circular search head made of metal or tough plastic. It can be saucer-sized or as big as a dinner plate.

While today's detectors are simple to understand and simple to use, you may have a bit of a problem selecting a machine that's perfectly suited to your needs. No one knows for sure how many people have

## The Equipment You Need   7

become so frustrated with equipment they've purchased that they've forsaken all forms of treasure hunting for all time, but the number must run well into the tens of thousands.

First, decide exactly what type of treasure hunting you'll be doing. Searching for coins requires a detector that's different from the one you need if you're going to be exploring for gold. There are big differences in

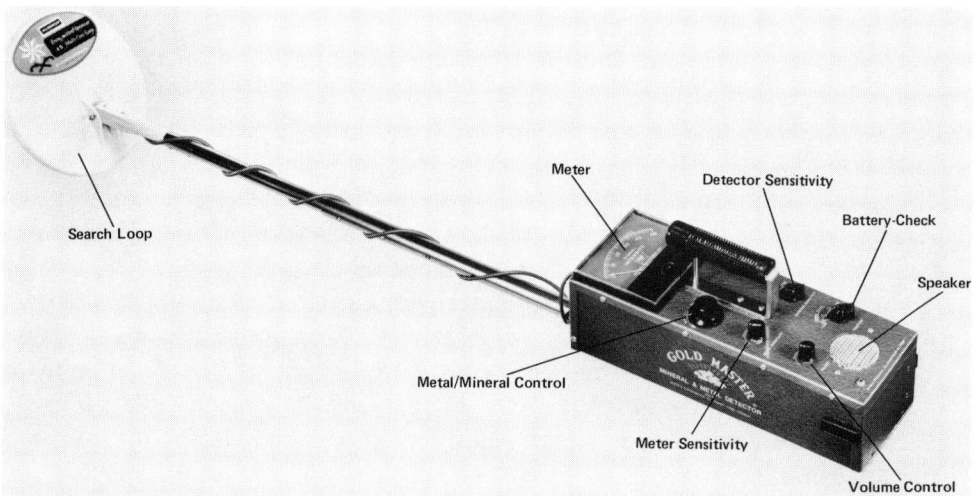

*(White's Electronics)*

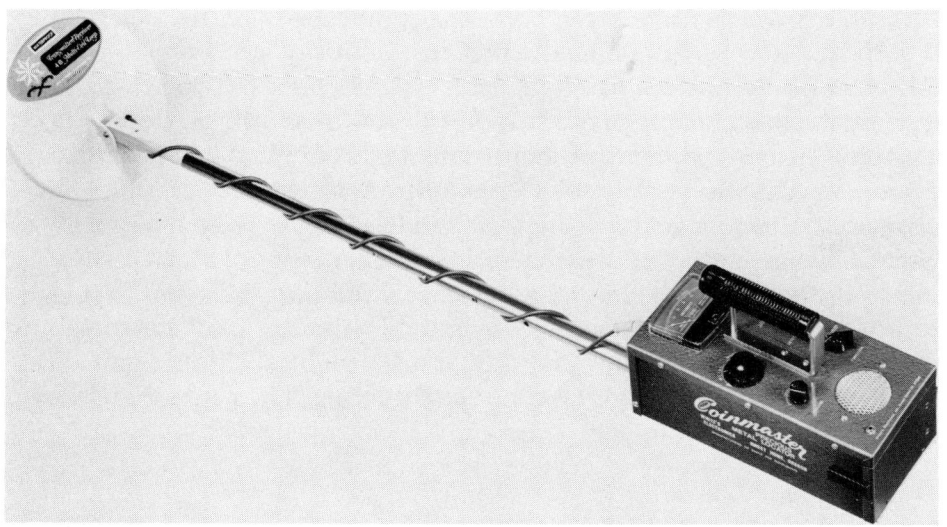

White's Coinmaster, a popular TR model. *(White's Electronics)*

search coils, too. Some are meant for deep-down detecting, while others are meant to perceive objects relatively close to the surface, within five or six inches of it.

Also bear in mind that not all detectors function effectively everywhere. Some perform erratically when used on saltwater beaches or where the soil is high in mineral content, while others are not bothered at all by such abnormal conditions. One thing to do is find out what make and model of detector is popular in your area. Talk to local treasure hunters. If you see someone using a detector in a park or on a beach, ask him what type it is and whether he recommends it. You can also check with the members of a local treasure-hunting club. Clubs are listed in the appendix of this book.

Appearance counts for a good deal when you're buying a detector. Select a machine that's ruggedly built, able to take some abuse. You want a unit that's going to give you many years of service.

The machine should have a comfortable "feel" to it when you test it out. This has to do with the way it is balanced. With some, you grip at the very end of the shaft, and thus your wrist and forearm are made to bear the full weight of the control box, the handle, the search coil—everything. Even if the unit happens to weigh relatively little, you're going to tire quickly using it.

Choose a detector in which the handle is so positioned that it enables the control box to counterbalance the weight of the search head. In other words, the handle will be at the point of equilibrium, or close to it. You'll be surprised how easy it is to use a machine of this type.

How much should you pay for a detector? It depends. If you're going to be using the machine merely to search for coins in a local park on summer weekends, you can probably find a machine that will perform up to your expectations for less than $100. But you have to be careful. Some low-priced detectors are no more than children's toys. They are able to perceive things like a Volkswagen at three inches, but not much else.

If you're going to be serious about treasure hunting, figure on paying from $150 to $250 for a detector. Most machines in this price range will give you sensitivity and dependability. It will also be a versatile unit, enabling you to do many types of treasure hunting. The pages that follow will help to guide you in your selection.

## TR *vs.* BFO

Before you make your first visit to a detector dealer, it's important to realize that there are, from the standpoint of electronics, two different types of detectors: transmitter receiver (TR) and beat frequency (BFO). You should know the basic characteristics of each.

The TR type, which has a direct link to the development work done by Dr. Fisher, is the more popular of the two, one reason being that you can operate it with greater facility. It is based upon the principle that a transmitted radio signal at a given frequency and amplitude will provide a uniform degree of amplitude—or tone—in a receiver, unless something interferes with that signal. Should the source of the interference that produces the tone change be a piece of metal—a coin, say—then in essence you have "detected" that object.

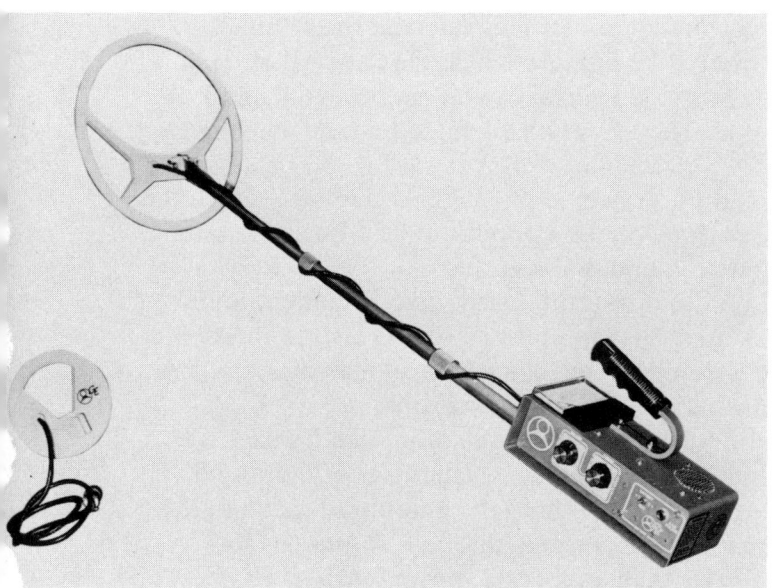

Garrett Electronics makes this BFO unit. *(Garrett Electronics)*

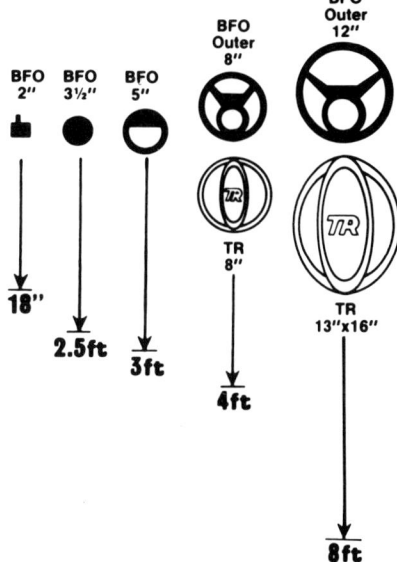

Search coils—the bigger the deeper. *(Garrett Electronics)*

When using a TR detector, the presence of metal is made known to you by a sudden loud and sustained high-pitched beep, or by means of a meter that abruptly registers the signal, or both.

The beat frequency detector is different in that it consists of two oscillators, each operating on the same frequency. One of these oscillators is variable and has an inductance coil that is generically known as a "search coil." When, through the presence of metal, a difference in frequency is produced between the two oscillators, the detector speaker and meter report the difference. But not in quite the same way as a TR unit does. When a BFO detector is tuned in properly, it emits a series of rapid beats; a "motor-boating" sound, manufacturers call it. When the search head is passed over anything made of metal, these beats increase in loudness and frequency.

BFO detectors date back to World War II when they were used by the military for detecting mines (not the kind that have anything to do with ore deposits). They used to be considered superior to the TR type because they could be used with interchangeable loops of different sizes. But in recent years TR models have included this feature. The subject of loops is covered in detail in the following section.

The differences in the way the TR and BFO machines perform is what's important. TR detectors are sometimes referred to as "quick response" machines. They are like light switches; they are either on or off. You get either a loud, sustained signal or you get no signal at all. If you're hunting for coins, this is no great problem. When the machine buzzes, you dig.

But TR machines don't differentiate—not yet. You get pretty much the same signal whether the detected metal happens to be a buried sewing thimble or the rusted fender of an old Chevrolet.

The failure of the TR unit to discriminate can be overcome to some extent by advancing the sound level so that you obtain a continuous signal. When something is detected, you then get a signal variation, and it's possible to learn to distinguish between variations.

The BFO detector is different. There is much more of a direct relationship between what is detected and the character of the sound. The beats increase in frequency in proportion to the amount of metal in the detectable field, until the machine's limit is reached. A quarter, for example, gives off fewer beats than an old monkey wrench at a comparable depth. It doesn't take long to learn to interpret the signal variations.

Another difference between TR and BFO machines is evident in those areas of the country where the soil happens to be heavily mineralized. In

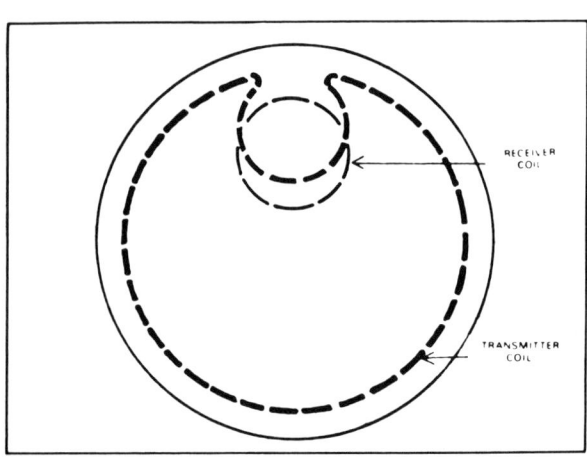

A two-coil TR unit; at first glance it seems like three. *(How-to-Test Detector Field Guide,* by Roy Lagal, Ram Publishing Co.)

the Rocky Mountain West, for example, where the ground is often pocked with deposits of black magnetic sand. If you use a TR machine over such ground, you will find the sand causes an audio response. Obviously, this is a problem. BFO units can be so tuned that they overcome this failing.

A fairly recent innovation is what is called the combination detector, one with both TR and BFO circuitry. With a flick of a switch, you choose the type you want.

The idea, of course, is to endow you with the best of both worlds. But it's moot whether this is actually the case. When a combination unit is tuned into the BFO mode of operation, it is capable of giving optimum performance. But when operating as a TR machine, there is a discernible loss of sensitivity.

To sum up, BFO machines are the more versatile, able to be used anywhere for all types of treasure hunting. They are a bit more difficult to use, in that they require that you learn to interpret the signal, not just perceive it. TR machines, characterized by their quick response, excel when coins are the object of your search. But watch out for a reduction in the quality of performance if you happen to encounter mineralized conditions.

## THE RIGHT SEARCH COIL

The matter of coils, or loops, as they're also called, is critical to your success. You have to choose the coil that's best suited for the type of treasure hunting you're going to be doing.

BFO coils are often of this design; TR coils are usually solid disks. *(George Sullivan)*

Regarding coils, keep these two facts uppermost in your mind:

*A small coil* (3 to 8 inches in diameter) detects objects of every size at shallow depths.

*A large coil* (8 to 24 inches in diameter) detects large objects at deeper depths.

Size is only part of it, however. You also have to differentiate between TR and BFO coils. As previously stated, in the case of a TR detector, one coil transmits the signal and the other receives it. So when the manufacturer of a TR unit advertises that his heads feature a dual-coil system, it is a rather meaningless boast. The head has to have two coils; it's the nature of the TR system. But only one coil—the receiving coil—actually responds to what is being detected.

Some other manufacturers of TR units offer what they call triple coils. A close examination may show the word "triple" to be something of a misnomer. The unit will have a receiver coil and a transmitter coil, but the latter is laid out in such a way that it only gives the appearance of a third coil. You can usually establish that there is only one working coil by field-testing the machine.

You encounter similar advertising claims in the case of BFO units. For instance, you may see "triplet" coils advertised, but a close examination will show that these are not three separate coils, but simply one coil wound in three connected concentric rings. So you actually have one search coil, and when the unit is in operation only one electromagnetic field is created.

Decide on the right size of coil first. A 6-inch loop will detect a dime buried at a depth of 4 inches more readily than a 10½ inch loop. Thus, coinshooters prefer smaller loops.

Larger loops are for detecting bigger objects that are deeply buried. If you're a bottle collector and are seeking a long-buried refuse pile, then a larger loop is definitely the one to use. there are specialized coils to use when prospecting for gold and others for deep-water work.

Some treasure hunters look for versatility when they're buying a detector. They may be probing relatively deep into the ground one day, and seeking small objects at shallow depths the next. No one coil is going to do both jobs.

But manufacturers have come up with two solutions to the problem: interchangeable coils and dual-coil units. With the former, you simply replace one coil with another, the way you replace a flat tire with a spare. But with coils it's much easier. It shouldn't take you any more than thirty seconds to make the changeover, even if you're not especially mechanically inclined.

Independently operated dual coils, common to some BFO units, are the second method of obtaining versatility. Developed during the 1960s, this system involves the use of a small coil that is mounted within the circumference of a larger one. In one of the Garrett Company's dual-coil machines, the inner coil is 5 inches in diameter, the outer one, 12 inches. A switch on the control panel enables you to select the one you want.

Since, in the case of BFO units, the coils are not encased in a molded covering, you can tell at a glance how many coils the unit has. But it's a good idea to test the coils to ascertain that each operates independently of the other. Simply lay the machine flat and pass a coin over the coils, first energizing one coil and then the other. The coil that is energized will cause a total response; the nonenergized coil will respond, too, but only slightly.

If you don't need versatility, don't bother with any of this. Simply buy the size of loop that's going to do the job you want. Search loops cost from $25 to $75, so when you buy one instead of two there's a big saving.

Search coils common to TR-type detectors vary widely in efficiency, depending on their design. I'm speaking here not about the quality of the electromagnetic field that is produced, but about the size of that field, its overall configuration. It is a matter of fact that in two coils of equal size, electromagnetic fields of different sizes can occur.

Remember, all TR units contain two-coil search loops. One is the transmitter coil that sends out the signal. A portion rebounds from the detected object to the second coil, the receiver coil. With many TR machines, the receiver coil is significantly smaller in circumference than the transmitter coil. With any TR detector you plan to buy, what you have to determine is whether the response area is limited to only the area covered by the receiver coil.

With some newer TR models, the receiver and transmitter coils are identical in size and mounted side by side within the loop housing (see diagram). Sometimes these are referred to as "total response" coils.

The total response-type assures you much greater efficiency in the field. Because there is responsiveness over the entire surface of the loop, you can cover a greater area with each sweep. On the other hand, when the response area is restricted in size, you have to overlap sweeps or you will be failing to cover big chunks of ground.

You can perform a simple experiment that will tell you whether the TR unit that you propose to buy has a total response coil. Tune in the machine so that it will respond to a small coin, then place the unit on a desk or table. Take the coin and beginning at the top of the loop, slide it in a zigzag pattern toward the bottom. You should get a continual response over the entire face of the coil. If, however, there is a big no-sound area from just above the coil center to the bottom, well, there is a problem.

Inspect BFO coils carefully, too. With some, there can be undue interference, a random crackling sound when the unit is used on wet grass or over mineralized ground. But this won't happen if the coils are properly electro-magnetically shielded.

Because a coil is rated as being waterproof, it doesn't mean that it has been shielded. A waterproof coil is one that can be operated underwater without any ill effect, but it still might cause static when used in wet grass.

No dealer is going to allow you to cut open the search loop to find out whether the coil has electrostatic insulation. So try this test: Tune in the detector so you get a moderate beat. Lay the detector down on a desk or table and take a handful of wet grass and rub it against the search loop. Notice whether there is any static or the signal is erratic in some fashion. Change the tuning slightly and repeat the test. Avoid tuning in the instrument so that the beat becomes exceptionally fast or slow.

You should also give some consideration as to how the coil is fixed to the shaft. Some swivel; others don't. With the former, you are able to adjust the coil angle, an advantage when scanning a high wall or searching a beam over your head.

A final word about coils: The capability of each depends on several factors besides the size of the loop and the other features mentioned here. The chemical composition of the soil is a factor and whether the ground is wet or not. How high the coil is held from the ground's surface is another important matter. Keep these facts in mind when field-testing any coil.

### DETECTOR ACCESSORIES

Some detectors are fitted out with so many accessory items that the control-box face looks like the instrument panel for a DC-10. With other machines, the stress is on simplicity. There's just one control switch; the unit works like a table lamp. The detector that you want undoubtedly falls somewhere in between these two extremes. The paragraphs that follow are meant to help you decide what's essential and what isn't.

*Tuning Control* It's not merely enough to be able to regulate the amount of sound you get from the speaker; the controls should enable you to fine tune it. Only through fine tuning will you be able to get the sensitivity you require for sophisticated searching.

*Metal-Mineral Tuner* This control knob enables you to select the type of material the search coil will detect. When coinshooting, for example, you turn it to the metal setting.

A dual coil detector. *(Garrett Electronics)*

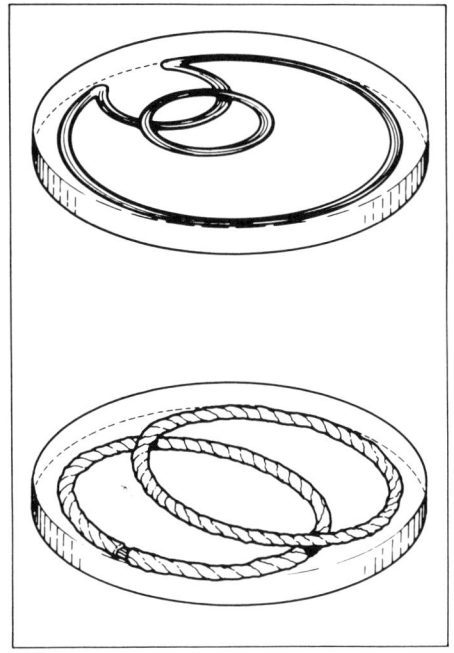

Narrow scan TR coil is shown at top, total response coil at the bottom. *(How-to-Test Detector Field Guide,* by Roy Lagal, Ram Publishing Co.)

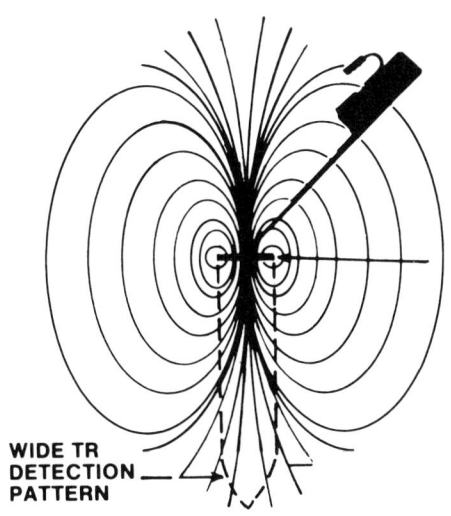

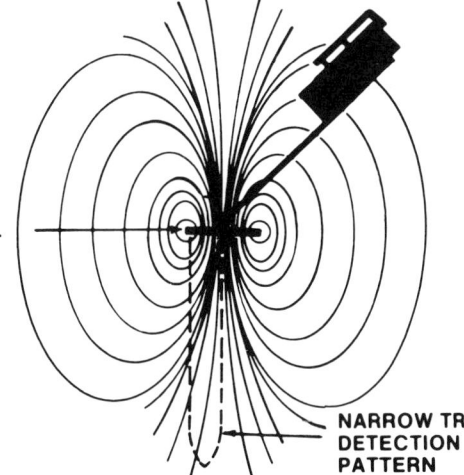

Narrow search pattern will sharply reduce the area you cover with each sweep. *(Garrett Electronics)*

Coin test will tell you whether coil gives "total response." *(George Sullivan)*

It must be said that it is somewhat misleading for detector manufacturers to claim that their machines are capable of distinguishing between "metal" objects and those that are "mineral." When tuned to the metal mode of operation, they do react to conductive metals: copper, brass, silver, or gold. But on the mineral setting, the detector *also* responds to magnetic minerals, specifically, to magnetic iron or iron oxide ($Fe_3O_4$). It so happens that most minerals are found where there are also concentrations of iron oxide.

There are an infinite number of real minerals that the machine won't respond to, no matter what setting you use. Try this test: Set the detector in the mineral setting and pass a piece of "pure" mineral under the search loop. Try something such as calcite, the basic constituent of limestone and marble. Notice that you don't get the slightest reading.

*View Meter*   This feature is known by several names. Sensitivity meter and intensity meter are two of them. What the meter does is give you a visual signal whenever metal (or magnetic iron) is perceived by the search loop. A strong signal indicates either a large object, or a small object very close to the surface. While the meter is more sensitive than the speaker, often reporting even the tiniest object, don't consider it an essential feature. Most treasure seekers use it only as a check on the audio system.

*Battery-Check Meter*   With this feature you're able to get an instant reading as to the operating condition of the detector's batteries. A switch on the control panel actuates the meter.

Batteries are of crucial importance, of course, and you can't always judge their condition by the manner in which your detector is performing. There are simply too many factors other than the batteries that affect the machine performance. With a battery-check meter, you're always sure exactly where you stand.

It's not easy to make a general statement as to the amount of use you can expect from the batteries that power your machine. Some manufacturers claim as much as 200 hours of battery life for their detectors, but most experienced treasure hunters consider this to be something of an exaggeration. Whatever the figure, it is very much related

It's simple to replace batteries. *(Garrett Electronics)*

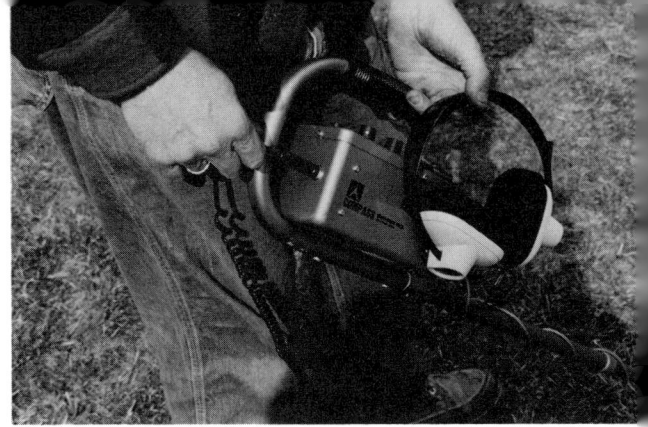

Most experienced detector operators use earphones. *(George Sullivan)*

to the number and type of accessories that the detector has. For example, earphones do not require as much power as a speaker system.

Many detectors now operate on penlight batteries, the small "AA" batteries, each about the size of a lipstick tube. Some require as many as 12 or 14 batteries of this size, and since they cost 30¢ to 35¢ apiece this many seem like quite an expense. But penlight batteries are superior to the single-battery system in that their hourly operating cost is less. Incidentally, batteries retired from use in detectors can still be used to power flashlights, portable radios, and small household appliances.

*Telescoping Shaft* The important word here is "telescoping." You should be able to adjust the length of the detector shaft to your particular height; otherwise, you're going to be uncomfortable and tire quickly as a result. And it's easier to store and transport a detector with a shaft that telescopes.

*Earphones* Some treasure hunters swear by earphones, saying that they're able to discern the slightest variations in the signal with greater clarity than when using the speaker. And it's true. Virtually all experienced treasure hunters use them. I'm talking here about *earphones,* a complete headset, not a mere earpiece.

Earphones are also valuable on windy days or in searching beach areas when the surf is high, that is, whenever you want to block out extraneous noise. More than a few operators use earphones as an excuse for not answering the questions of the curious. When someone asks, "What are you looking for?" standard procedure is to point to the headset and shake your head to indicate a lack of comprehension. Earphones cost from $10 to $25.

*Discriminator* From the time that metal-mineral detectors first came into widespread use, one problem has been a lack of selectivity. *Anything* made of metal starts them chirping. This means that along with coins and rings and the like, you're also going to "discover" countless bottle caps, metal pulltabs from soda and beer cans, the cans themselves, pieces of aluminum foil, and many other and varied metallic items that can only be classified as junk.

Some manufacturers have recently begun to offer detectors that are meant to solve this problem. These machines have what is called a "discriminator" feature; they are able to distinguish the worthwhile from the worthless.

Most units of this type employ both a meter system and a speaker. When a coin is detected, the meter needle swings to the right and the sound becomes louder. In the presence of an unwanted object, again the sound gets louder but this time the meter needle swings left. Garrett Electronics, Pacific Northwest Instruments (Bounty Hunter), and Gardiner Electronics Company are among the firms that market machines of this type.

Machines with the discriminator feature are best suited for use on beaches or in parks when the user is coinshooting. It's then that junk items cause the most grief. You wouldn't want to use one when investigating a ghost town or an abandoned building because the machine would probably reject relics and other objects that are worthwhile.

There is much debate among treasure seekers as to the real value of detectors of this type. Many question their sensitivity. Most veteran treasure hunters prefer to rely on their own sharply honed skills. When you become experienced in the use of a detector, you learn to "read" signals as they're emitted by the speaker or earphones. There is a subtle difference between what you hear when the search loop passes over a bottle cap and what you hear when a dime is detected. In other words, a skilled operator endows his machine with selectivity.

*Waterproof Search Loops* "Waterproof" is a word often used carelessly. It should mean impenetrable to or unaffected by water. When the search loop is properly waterproofed, you'll be able to submerge it and the detector shaft as far up the shaft as the control box. Some manufacturers make an additional charge for waterproof loops; with others, they're standard equipment. Waterproof should also mean that the coil connections—those within the coil face—are not going to be affected by water either.

*Coil Covers* Available to fit search loops of several different sizes— 6-inch, 8-inch, and 12-inch—these covers snap in place to protect the loop from wear and damage when being used over sand, gravel, or rock-strewn terrain. They're also recommended when working on wet grass when your search loop happens to be only "water resistant," not waterproof. Coil covers are made by several different firms. They cost $3 to $5, depending on the size.

*Speed Handle* An auxiliary handle that attaches to the detector shaft, this device permits two-hand control and, thus, is said to enable the user to work for longer periods with a minimum of fatigue. It also increases the range of each sweep, permitting one to make a full 180-degree arc.

The speed handle can be used with detectors of virtually every type (not solely body-mount instruments, as pictured here). Manufactured by Compass Electronics, the speed handle costs $12.

*Indicator Light* At least one detector can be used after dark. The Yukon 99-B (Compass Electronics), is a machine with an indicator light that signals a find with a soft, green glow.

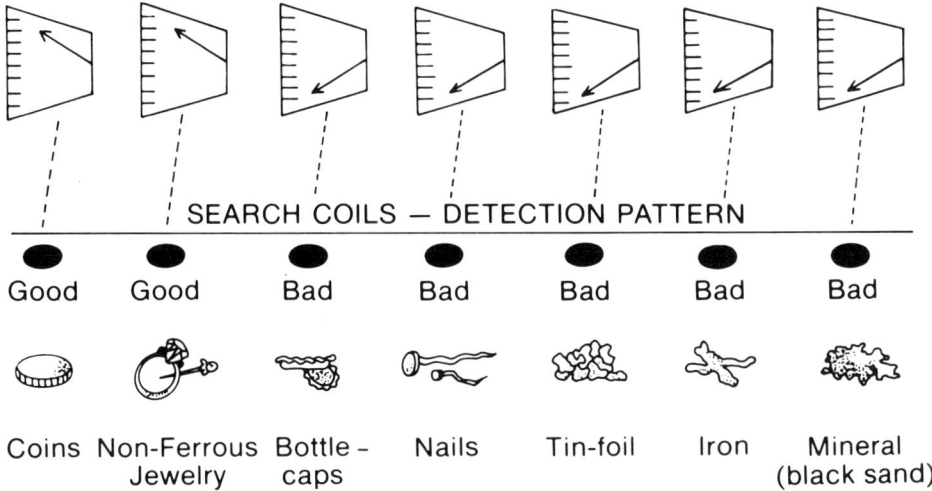

Discriminator enables the operator to distinguish "bad" objects from "good" ones. *(Garrett Electronics)*

## DETECTOR DEALERS AND MANUFACTURERS

There are probably half a dozen or so detector dealers in your area, each representing one or more of the country's twenty to thirty manufacturing firms. Dealers are listed in your Yellow Pages under "Metal Locating Equipment."

You don't have to buy from a dealer; you can order a detector by mail from the manufacturer, but it's better to buy from a dealer. Then you have a chance to try out several different models and get a "feel" of each.

Buying from a dealer is also likely to be an advantage when it comes to getting your detector serviced. Virtually all manufacturers permit you to send a detector to the factory for repairs or when new parts are needed. But this presents problems. Not only do you have to pack up the unit and pay the postage, but there can be considerable waiting time involved. You're much better off buying from someone who is equipped to perform service chores. In such a case, it's not likely you'll ever be without a machine for more than 72 hours.

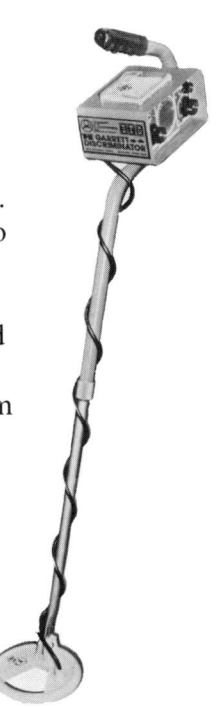

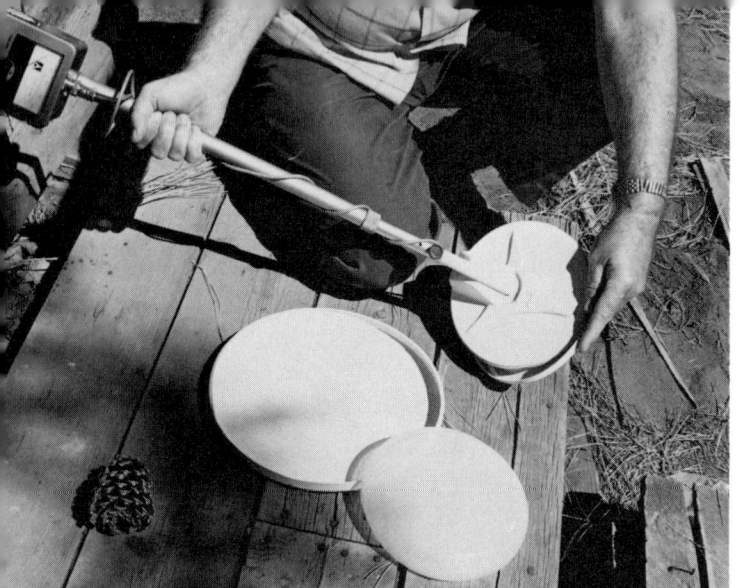

Coil covers. *(Compass Electronics Corp.)*

Using a speed handle. *(Compass Electronics Corp.)*

Be sure to protect yourself by getting a guarantee. Parts should be warranted to be free from defects for 24 months following the date of purchase. Within that time period, the company should repair or replace any defective part, or make any necessary repairs without charge to you for either the parts or labor.

There is not much of an opportunity to buy secondhand equipment, although you can check with your local dealer to find out whether he might have available any detectors he's received as trade-ins. You can also write to White's Electronics (1011 Pleasant Valley Rd., Sweet Home, Oregon 97386) and request information on detectors that their salesmen have used as demonstrators. They'll send you a list of models available and the prices for each. Address your request to the Used Equipment Department.

Listed below are the companies that manufacture and sell electronic detectors in the United States. (Those companies marked with an asterisk are major firms, each with a dealer network.) Write to several firms and request a catalog from each and also the names of dealers in your area.

*Bounty Hunter, Pacific Northwest Instruments, 3104 Johns Ave., Klamath Falls, OR 97601.
*Compass Electronics Corp., 3700 24th Ave., P.O. Box 366, Forest Grove, OR 97166.
*Detectron, Division of Tinker & Rasor, 417 Agostino Rd., San Gabriel, CA 91778.
D-Tex Electronics, 614 Easy St., Garland, TX 75040.
Excelsior Electronics, 7448 Deering Ave., Canoga Park, CA 91303.
Fisher Research Laboratory, P.O. Box 490, Belmont, CA 94002.

Gardiner Electronics Co., 4729 N. 7th Ave., Phoenix, AZ 85013.
*Garrett Electronics, 2814 National Drive, Garland, TX 75041.
General Electronic Detection Co., 16238 Lakewood Blvd., Bellflower, CA 90706.
The Goldak Co., Inc., 1101-A Air Way, Glendale, CA 91201.
Jetco Electronic Industries, 1133 Barranca Drive, El Paso, TX 79935.
Metrotech Underground Explorations, Box 793, Menlo Park, CA 94025.
Rainbow Manufacturing Co., 428 Louisiana Blvd., SE, Albuquerque, NM 87108.
Ray Jefferson, Main and Cotton Sts., Philadelphia, PA 19127.
Relco Industries, P.O. Box 10839, Houston, TX 77018.
Search Electronics, P.O. Box 8396, St. Petersburg, FL 33738.
Sherando Corporation, 101 Kings Highway East, Haddonfield, NJ 08033.
Solidtronics, Tennent, NJ 07763.
Treasure Electronics, Route 1, Box 56, Benton, LA 71006.
*White's Electronics Inc., 1011 Pleasant Valley Rd., Sweet Home, OR 97386.
Yosemite Electronics, 1950 Divisadero, Fresno, CA 93701.

These firms manufacture more than two hundred different detector models. Some of the more highly regarded units are described in the pages that follow.

*Gold Master S-63 Deluxe (White's Electronics Inc.; TR)* One of the most popular of all detectors, this unit is the favorite of countless coinshooters. Powered by 14 1½-volt penlight batteries, the Gold Master is equipped with a waterproof 7½-inch search loop. The unit weighs 4 pounds, 14 ounces, which may seem heavy, but it is balanced so well that the weight is not a factor. The machine reports finds audibly and also by means of a speaker system. The meter also includes a battery-check feature. There's a headphone jack but the earpiece supplied with the unit in not of great value. Overall, the Gold Master is a sensitive, stable instrument, sturdily built. It is guaranteed for two years. List price: $179.90.

To extend the versatility of the Gold Master, you can purchase a second loop, a 10½-inch loop. This enables you to detect large objects at greater depths. It costs: $69.50

*Master Hunter (Garrett; BFO)* Designed to perform every treasure-hunting job, this machine comes equipped with two waterproof search heads, each with two search coils. One has 3½- and 8-inch coils; the other, 5½- and 12-inch coils. Once you decide on which head to use, you switch on the appropriate coil. The unit offers the company's "zero-drift" circuitry, the new "Triple-Output" speaker, and Faraday shielded coils.

Except for the view meter, the controls are mounted conveniently on the (recessed) left-hand panel of the control console. The unit is also equipped with a battery-test meter and an easily operated single knob tuning system. It is powered by 3 9-volt batteries and the cost of operation is estimated to be 5 cents an hour. The Master Hunter weighs four pounds. List price: $249.50. Headphones are an additional $12.50.

Garrett's Coin Hunter, another BFO machine, is essentially the same unit, but it is equipped with only one dual-coil search head, the one with the 3½-inch and 8-inch search coils. The larger coil enables you to probe deep into the ground and cover wide swathes with each sweep. By switching to the 3½-inch coil, you pinpoint the find. List price: $199.95.

*Yukon 77-B (Compass Electronics; TR)* A deluxe instrument with a wide-scanning, deep-penetrating 8-inch waterproof search loop, the Yukon 77-B offers a full range of signal controls. A ground control tunes out background interference. There is also a fine-tuning control to obtain greater signal sensitivity, a meter control to help confirm finds, as well as a volume control, battery check, and headset jack. The machine weighs less than four pounds and gets high marks for balance. List price: $219.50.

With a second loop, a 12-inch loop, the Yukon 77-B is priced at $264.50.

The Yukon 77-B can be adapted for use with a body mount. Herein the control console is fixed to the user's belt and linked with the search loop and its shaft by a long cord. The operator thus has far more control over the loop, and is able to probe rugged terrain or hard-to-get places with greater ease. Rigged as a body-mount instrument and equipped with an 8-inch loop, the unit lists at $219.50. Add a 5-inch loop to the package and the price is $264.50.

*Master Hunter (Garrett; TR)* Probably the most versatile of all TR units, this machine comes with two search heads, one with an 8-inch coil, the other with a 13-inch *and* 16-inch coil. Both heads boast Garrett's "wide scan" feature, both are waterproof, and electro-magnetically shielded. The Master Hunter offers both meter and speaker response and a battery test meter. Like Garrett's BFO machine, this unit is powered by 3 9-volt batteries; operating costs are estimated at 5 cents an hour. List price: $249.95, with an additional $12.50 for earphones.

*Bounty Hunter III (Pacific Northwest Instruments, Inc.; BFO)* A sensitive, reliable machine. Offers vernier tuner, battery-check system, powerful speaker, and an easily read meter. With shielded, waterproof 6-inch coil, list price is $129.95.

*The Beachcomber (White's Electronics, Inc.; TR)* A relatively new model, weighing only 42 ounces and operating on 6 penlight batteries, the Beachcomber is equipped with an 8-inch search loop that is fixed permanently to the shaft; in other words you can't switch to an alternate loop. Ease of operation is one of the Beachcomber's chief features, since there are only two control knobs, a volume control and tuner control, the

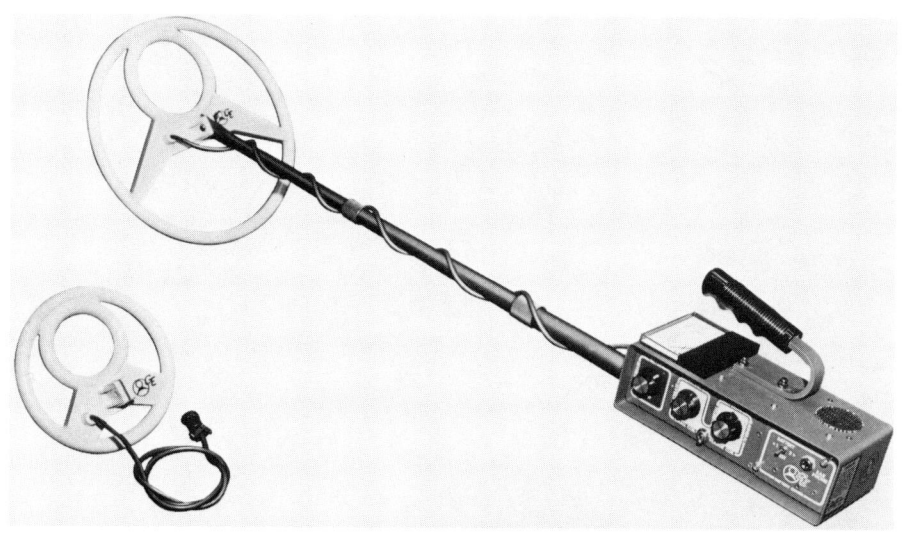

The BFO Master Hunter.
*(Garrett Electronics)*

The Yukon 77B, with 8– and 12-inch loops, earphones.
*(Compass Electronics)*

The TR Master Hunter.
*(Garrett Electronics)*

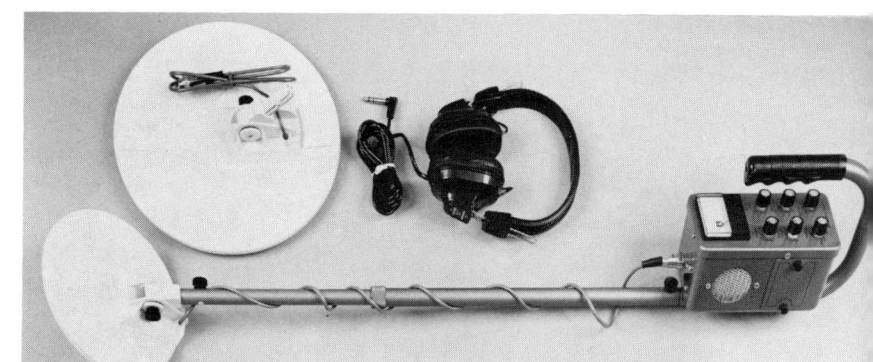

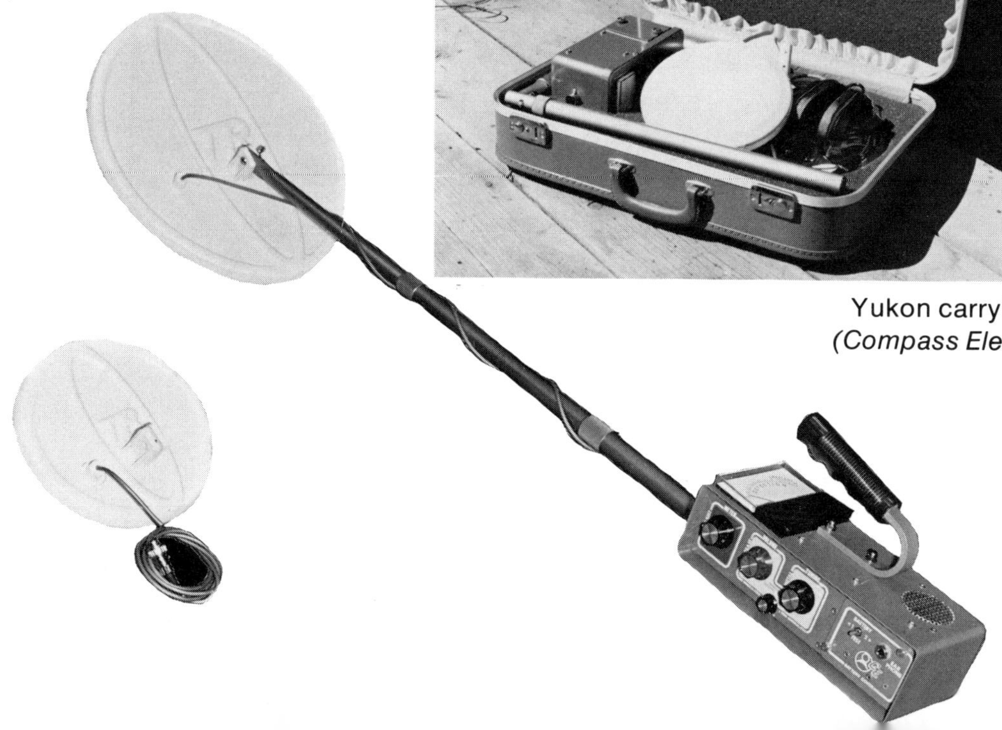

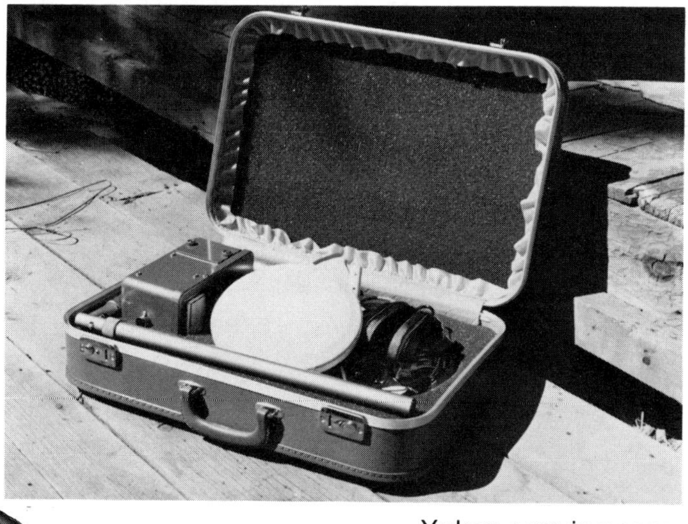

Yukon carrying case.
*(Compass Electronics)*

The Beachcomber.
*(White's Electronics)*

Detectron Model 7-T.
*(Detectron)*

The Myty-Myte.
*(George Sullivan)*

latter incorporating the power switch. The unit is equipped with an earphone jack. Because of its light weight and ease of operation, many dealers recommend the Beachcomber for novices or children. But it's no toy. The instrument will detect gold, copper, silver, platinum, and other metals through mud, dirt, snow, ice, brick, and rock. List price: $99.95.

*Detectron Model 7-T (Detectron; TR)* Light in weight (2¾ pounds) and low in price ($139.50), this machine features a single-knob control system and a rugged "Cycolac" search head, 8 inches in diameter. It also offers a battery-test meter and transistorized circuitry. It operates on 3 "AA" penlight cells; warranted for one year.

*Myty-Myte (Excelsior Electronics Co.; TR)* Economy and simplicity of operation are the chief features of this unit. It's easier to use than a

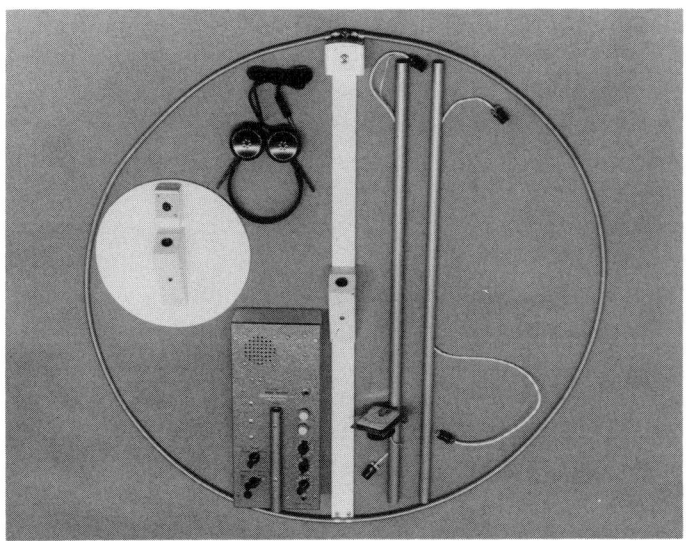

Gardiner 181-B offers 36-inch search coil (outer ring).

portable radio. You switch it on, then adjust the speaker volume control to the threshold level; that's all. Equipped with a 6-inch waterproof search coil, it can detect a dime at 6 inches. Telescopic handle extends to within an inch of 4 feet. Two-year warrantee against defective parts or workmanship. Price, $39.95; with earphones, $49.95.

*Model 181B (Gardiner; TR)* This is a sophisticated unit, able to distinguish between small metal objects of various types. When the search coil is passed over an iron object, such as a nail or bottle cap, the meter needle deflects to the left and the signal pitch goes lower; when passed over a noniron metallic item—a dime, for example—the needle deflects to the right and the pitch becomes higher.

The machine comes equipped with two search coils, a conventional 11-inch coil and a quite unconventional 36-inch one. What can you detect with a search coil that's 3 feet in diameter? A flashlight case at 12 inches, a quart can at 36 inches, or a 5-gallon can at 56 inches. Its maximum range is given as 17 feet.

The 181B is powered by 6 AA-size penlight batteries and has a battery-check meter. Its control system enables you to eliminate background interference and you can fine-tune the machine. Headphones are included at no additional charge.

The unit weighs about 4 pounds. It offers a longer-than-normal shaft, with the control box mounted at shaft end. The operator grips the shaft when using the machine, taking his grip at the balance point or near it. List price: $385.00.

For locating large detectable objects at depths of 5 to 10 feet or even beyond, what's needed is an unusual type of TR machine, one referred to as a "double box" detector. The search coils are contained in a metal case at one end of the unit, and the signal they transmit is received from the ground by a receiver that is located at the other end, where the search loop is usually found. The presence of ore veins, an abandoned mining tunnel (with its detectable metal rails and air ducts), a cache of buried coins or whatever, is reported to the operator by means of a meter, or audibly via earphones. There's no speaker system.

Utility companies use these deep-seeking detectors for locating sewer lines or gas lines, and they are also used to find buried survey stakes. But for small objects, anything smaller than a belt buckle, say, they are not of great value.

Several companies manufacture double box units, including White's Electronics (with its Super X-500), Detectron (711-T), Metrotech (330 Eldorado), Fisher (Gemini), and Goldak (Bonanza). Prices for these machines are about the same as for the more conventional detectors.

## TRIAL RUNS

Not all detector manufacturers are noted for their ethical conduct, and you will likely see advertising claims that, if not dishonest, are at least questionable. Be sure the detector you plan to buy performs as the dealer says it will. Put it through a series of tests before making your final decision.

Drift, the tendency for a detector's signal to wander from its set position, is a problem with lower quality instruments. When the signal

For deep searching—5- to 10-feet—a double box detector is often used. This is Goldak's Bonanza. *(Goldak Corporation)*

either climbs abruptly to a high pitch or quickly reduces in pitch—for no apparent reason—that's drift. The problem can occur in both TR and BFO machines.

If you don't permit your detector to warm up sufficiently, or if, in the field, you encounter a change in ground conditions (from wet to dry, or from nonmineral to mineral soil), you're likely to experience some drift. But this is to be expected. The type you shouldn't anticipate is that which occurs spontaneously because of some design failing or other inadequacy.

Before you buy a detector, test it for drift. Allow the machine to warm up, giving it at least ten minutes. Try out the detector in the store, listening carefully to its signal, and then ask to take it outside. A change in temperature, going from an air-conditioned store to the warm temperature outside (or from a heated store to a colder outside temperature) will cause drift—if the machine is susceptible to it. Do this test yourself. If you allow the dealer to do it, he may adjust the instruments so that the drift condition will be masked.

Besides being drift-free, you should also assure that the signal you get is stable, that is, one that is not going to be distorted by abnormal fluctuations that the machine itself creates. When you field-test the unit, hit the sides of the control box with the heel of your hand a couple of times. The detector should be able to withstand such treatment without any change in the signal pitch or frequency. Any lack of stability is an indication of inferior quality.

Stability should always characterize the signal as you sweep the search loop from side to side. With a poor machine the signal may grow louder just as you end one sweep and prepare to begin the next. The trouble such false signals can cause is obvious.

Another way to test for stability is to raise and then quickly lower the search head several times, bringing it to a point high above your head on each swing. Perform this test outside, well away from any objects that might cause interference. There should be no change in the signal as you raise and lower the loop.

Also test the machine for sensitivity. Lay the detector flat on a desk or on the dealer's showcase, and tune it in. Use the metal mode of operation. If the dealer says the machine is capable of detecting a dime at eight inches, hold a dime that distance from the search loop and see whether you get a response.

An even better way to test for sensitivity is to bury coins and other objects at various depths and see whether you get a response from them. Keep in mind that it is more difficult to get a response from freshly buried coins than it is from those that have been in the ground for some time.

Metal buried for several years oxidizes and in so doing creates a detectable field in the surrounding soil. A machine that will respond to a freshly buried cent at five or six inches should be able to detect a cent in the ground for four or five years at depths of seven or eight inches.

But perhaps you're not planning on coinshooting. If that's the case, then field-test the detector with whatever type of object you plan to search for. Use a search loop of appropriate size, a 12-inch loop, say, if you're going to be seeking deeply buried objects. Keep in mind that when testing detectors made by different manufacturers, you must use search coils of the same size in order to get valid results.

The topic of detector selection and trial runs is covered in detail in Roy Lagal's handbook, *Detector Field Guide, How to Test.* It's available for $3 from Ram Publishing Co. (P.O. Box 38464, Dallas, TX 75238).

## BUILDING YOUR OWN DETECTOR

Electronic equipment of just about every kind is available in kit form today, and the roster includes metal detectors, or as Heathkit, the largest company in the field calls them, metal locators. The reason for buying a kit and doing your own assembly work is, of course, to save money. Figure the saving to be $30 to $40 on a detector that would have a retail price of $150.

Heathkit offers two detector kits. There's the GD-348, which employs a transmitter receiver system, the signal controlled by a ten-position hand-adjusted screw knob. The unit can also be used with earphones. It features a pistol grip handle and a telescoping shaft. Handle, shaft, the search coil, and control box fold into an easily portable package. A 9-volt battery supplies the power.

The kit is priced at $99.95. A carrying case for the unit costs $10.45. Heathkit says that it will take you three evenings to do the assembly work.

Heathkit also affers an economy model, its GD-48, which is similar in design and operation to the GD-348. It costs $74.95.

For additional information on these kits, obtain a free catalog from a local Heathkit sales and service center, or write to the headquarters of the Heath Company (Benton Harbor, MI 49022) and request one.

If you're undecided about purchasing a kit because you think it might be too difficult to assemble, you can obtain the kit instruction manual and look it over. Heathkit sells the manual for either of the kits mentioned above for $2. Later, if you decide to buy the kit, the $2 is refunded.

The Radio Shack (2617 West 7th St., Fort Worth, TX 76107), with more than two thousand stores throughout the United States and Canada, offers two metal detector kits, one priced at $34.95, the other at $19.95.

They're available at most outlets or you can buy by mail. The company will mail free catalogs.

If you happen to be very skilled as an electronics technician, you may want to build a detector from scratch, buying your own components and assembling them. Detailed plans for doing this are available from these sources:

*Radio Electronics.* November 1967 (pp. 35-37), "Build a Treasure Finder." This detector's principle of operation involves a turned-loop oscillator and a crystal filter that act in combination to provide sensing and indicating signals. "It is," says the text, "simple, stable, and sensitive and easy to build and operate."

When the search loop is brought close to a metal object—coins lost on a beach, keys dropped in a snowbank—eddy currents are induced that serve to decrease loop induction, changing the oscillating frequency. The result is that the crystal filter passes a greater amount of energy to the intensity meter, whose needle the operator keeps reading.

The text, by Charles D. Rakes, covers three pages. There are seven photos and diagrams.

*Popular Electronics.* January 1967 (pp. 41-48, 94-96), "IC-67 Metal Locator." This instrument is what the magazine calls a "deep type" detector, one that has its greatest value in spotting pipes, cable, water irrigation valves, and the like at depths of up to seven or eight feet. It's not for coins, rings, and other small objects located within a few inches of the surface.

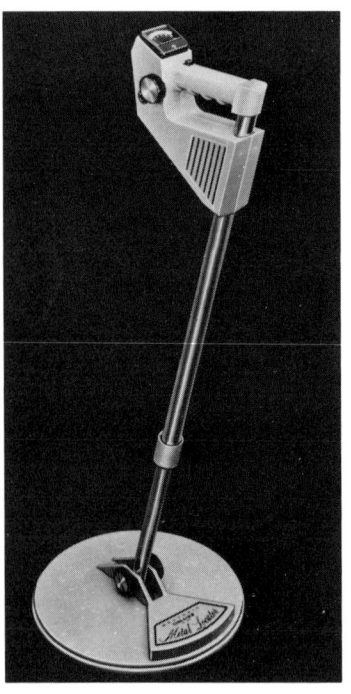
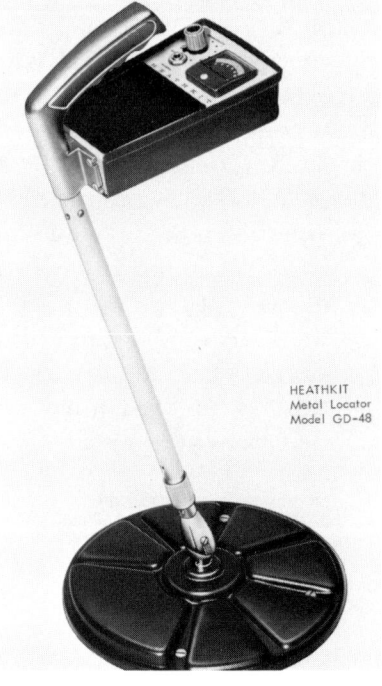

Heathkit offers these two models. *(Heathkit)*

The unit operates on the receiver-transmitter principle. It employs solid-state integrated circuitry and is powered by six penlight cells.

Some of the features it offers are not found in commercial units. Background noises and signal variations can be eliminated to permit the user to concentrate on the object of the search. The unit's control system enables the operator to pinpoint the location of buried objects sharply.

The IC-67 weighs four pounds. Materials to build it cost between $40 and $50, or about one-fourth of the amount you'd pay for a similar locator if purchased on the retail market.

"While not intended as a beginner's project," says the article, "the circuit is not too difficult to build and the parts are easy to get." Sixteen drawings and photographs help to amplify the thirteen pages of text.

*Popular Electronics.* July 1967 (pp. 27-32; 84), "The Beachcomber—Metal Locator for Treasure Hunters." Here is a simple, single-loop beat-frequency detector, described as "easy to assemble," and which is meant to find small objects only one or two inches in diameter within two feet of the surface. It will operate from six to eight hours on its transistor radio battery. Cost: about $15. Twelve photos and diagrams accompany the six pages of text.

Smith, Alan. *Introduction to Treasure Hunting.* Harrisburg, PA: Stackpole Books, 1971, $5.95. "How to Make Your Own Detector" is the title of a chapter in this book, and the detector it describes is a beat-frequency unit, one meant to perceive small objects at depths of up to eight to ten inches. The cost of the unit, says the text, shouldn't exceed $40. "The result will be a quality instrument well able to compete with commercial instruments in the $150 class."

Two construction approaches are detailed: You can either hand wire the chassis or utilize a printed circuit. Tools, supplies, electronic parts, and other hardware are described in carefully prepared lists. In total, the instructions cover twenty-four pages. Eleven diagrams and photos are included.

*Scientific American.* February 1968 (pp. 124-128), "The Amateur Scientist: Building a Sensitive Magnetometer and an Accurate Solid-State Timer." Compared to some of the sophisticated instruments available from major detector manufacturers, the unit described in this article is quite primitive but it is just what is needed for demonstrating the scientific principles that underlie these machines. It costs only a few dollars to build.

In a strict sense, the magnetometer is an instrument used for detecting the intensity and direction of magnetic fields. Commercial magnetometers are often used by surveyors to find buried iron survey stakes, or a utility company might use one to find buried iron pipe or conduits. In other words, magnetometers react to iron and steel—and nothing else. They can probe as deep as three or four feet.

The magnetometer described in this article is based on the effect a magnetic field exerts on protons to be found in the nuclei of atoms of water. The essential parts of the instrument are two plastic bottles wrapped with insulated copper wire and filled with distilled water. The bottles are fixed at right angles to the ends of a nine-foot pole, upon which is mounted a three-volt dry battery to power the instrument, a switch, and an amplifier which is linked to a pair of earphones. If the magnetic field at one bottle is different from that at the other, the induced frequencies also differ and a net difference in voltage is transmitted to the terminals of the amplifier.

You're not going to find anything like a survey stake at four feet with such an instrument. It compares to a commercial magnetometer in somewhat the same way a radio receiver using a crystal detector compares to the latest model FM receiver off the Sony assembly line. But like the crystal set, this simple magnetometer can provide exciting fun.

## DIVINING RODS

Publications in the treasure-hunting field, and also those such as *Popular Science*, carry advertisements not only for manufacturers of detectors and related equipment, but also for firms that sell divination instruments. Such devices have been used for centuries in searching for buried treasure and also for underground water reserves. When Marco Polo traveled to China in the thirteenth century, he reported on the use of these instruments to locate objects of different types.

Divining instruments come in a number of shapes and sizes. The forked stick is the best known. Used to locate the best spot to dig a well, it is supposed to be drawn to water in much the same way a magnet is drawn to steel. The user grasps the forked end and holds the stick parallel to the ground. When he passes over water, the end of the stick is supposed to point toward the ground.

L-shaped rods made of coat-hanger wire have been used to locate underground pipes. Another type of divining instrument takes the form of a short length of highly polished chrome, which is placed on one's outstretched palm. When the object of the search is perceived, the instrument tumbles out of the user's hand to the ground.

No matter what type of instrument he uses, the diviner is said to enter an altered state of consciousness. Through the instrument, he is said to derive data from some inner source, one that provides information

*The Equipment You Need* 33

beyond the usual sensory output. At least that's what specialists in the art say, and I do not dispute them.

One way to test your talent in the use of a divining rod is with this simple experiment. All you need is a wire coat hanger and a garden hose. Untwist the hanger until you have a straight length of wire. At a point about six inches from one end, bend the wire to form a 90-degree angle. This is your handle.

Stretch the garden hose across an open area and start the water running. Hold the rod loosely in one hand in front of your body with the longer section pointing directly toward the ground. Station yourself about twenty feet from the hose and begin walking toward it. As you near the hose and cross over it, the rod should revolve so as to point toward it.

Divining rods dates ancient times; this is from a 16th-century woodcut. *(New York Public Library)*

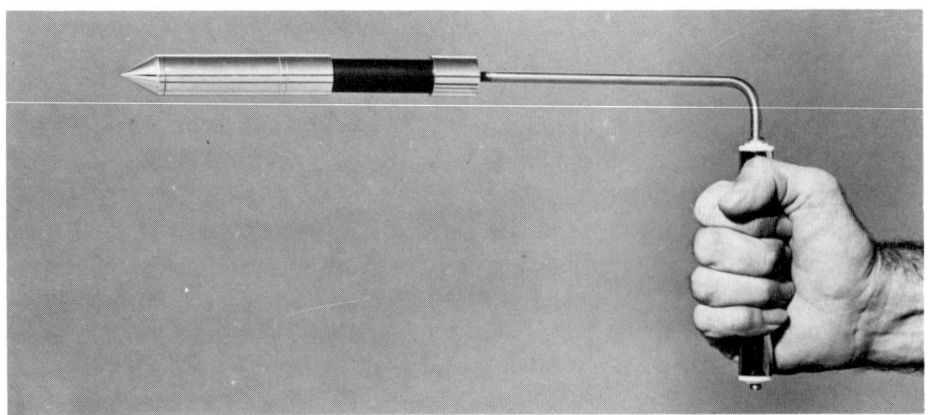

A modern-day "mineral rod." *(Carl Anderson)*

Treasure hunters are offered divining instruments that are said to be a great deal more sophisticated than bent hangers. The Spanish Dip Needle, also known as the Mexican Dip Needle, is said to have a range of up to twenty miles when seeking "large treasures or precious metal deposits." The Titan Mineral Rod, an L-shaped instrument, "can follow an attraction as deep as forty feet and can pull to a single coin up to fifty feet away, says the manufacturer. These devices are not inexpensive. The Titan Metal Rod costs $140.

You can obtain information on these and other instruments by writing for the free catalog from the manufacturer, Carl Anderson, Box 13441, Tampa, FL 33611.

After reading this chapter and consulting with detector dealers in your area, it may be that you are unable to find a detector that is absolutely perfect for your needs. Take consolation in the fact that success in treasure hunting is not dependent solely on the detector. You won't have good results with shoddy equipment, of course; quality is vital. But differences in performance between detectors of similar makes and models often derive from the manner in which the equipment is operated—the subject of the next chapter.

Chapter 2

# HOW TO USE IT

The first time, or even the first few times, you use a detector in the field you are likely to feel conspicuous. "You think everyone around is watching you," says a Long Island housewife. "The first time I went out with a machine, I wore a kerchief around my head so nobody I knew would recognize me."

It may help you to realize that this feeling of self-consciousness is very common at first. It applies not only to treasure hunting but every other outdoor activity. If you're a golfer, you know how ill-at-ease you were the first time you stepped up to tee off. You felt like every person on the course had his eyes glued on you. Tennis players confess to being uncomfortable the first time they go onto the court.

With perseverance you overcome. As you become skilled in using the detector, your confidence grows. After a while it doesn't make any difference whether one person or a dozen happen to be watching you. "Everyone is doing his or her thing today; that's the way I look at it," says a treasure-hunting veteran. "Well, using a detector is my thing. I don't care if it looks like Arnie's Army is following me."

Of course, the matter of becoming skilled takes time. It's not likely you'll get optimum performance from the machine until you've used it several times, particularly if the control system is complex. But each time you take it out, you'll learn something.

You can reduce the length of the period of orientation by practicing at home. One thing to practice is tuning.

### TUNING THE DETECTOR

Tuning is an art. If you undertune or overtune the machine, you can't expect to be successful with it.

The secret of tuning correctly is to have a delicate touch as you adjust the control knob. In the case of a TR detector, rotate the knob until you reach a point where there is no tone coming from the speaker and the meter needle rests on zero. (If your machine has both a mineral and metal setting, use the metal mode of operation.) Now continue turning the knob slowly until you first hear sound and the needle begins to move. Then slowly turn the tuning knob in the other direction; keep turning until the tone dies away and the meter again reads zero.

Put a coin under a magazine or book and sweep the search coil over it. You should now get a sound and meter reading when the coil passes over the coin.

When tuned in this fashion, the detector operates something like a flashlight; that is, it's either off (silent) or on (emitting a signal). This type of tuning is adequate for most types of searching, but for greater sensitivity you should learn to operate the unit when the sound is adjusted to what is called the threshold level.

Begin again at a point where there is no sound being emitted and the meter is at zero. Turn the tuning knob slowly until you first hear sound and the needle begins to move.

Now sweep the search coil over the concealed coin. Notice the abrupt increase in the intensity of the signal as the coil passes over the coin.

Try each method several times, varying the distance between the search loop and the coin. Notice that when you operate without sound, there are times that the machine fails to detect the coin. But when you use continuous sound, the machine's performance level is much improved. You must, however, be at that threshold level, the sound barely audible. If you overtune and get too much sound, you won't be able to perceive all the signal variations.

Veteran treasure hunters will tell you that there is a 20-30 percent increase in efficiency when you use continuous sound with a TR unit. "You not only are able to get a greater number of signals," says one, "but you get signal variations that enable you to get some idea of what type of object is being detected."

Think of the meter, which responds in concert with the speaker, as a backup system, confirming what the speaker reports. Of course, you can use the meter by itself, tuning out the sound. But this is awkward to do. When you're searching a piece of ground, it's natural to look at the search loop as you sweep. Keeping an eye on the meter slows you down.

Some makes of detectors can be further tuned by means of a range control setting, which is used in areas where meter variations tend to be excessive. Some machines manufactured by White's, for example, have a range control with four different positions. In the No. 1 position, the meter is at its most sensitive; at No. 2, it is half as sensitive; at No. 3, a third as sensitive, and at No. 4, a fourth as sensitive. Remember, it's not the sensitivity of the coil that's affected, only the meter circuitry.

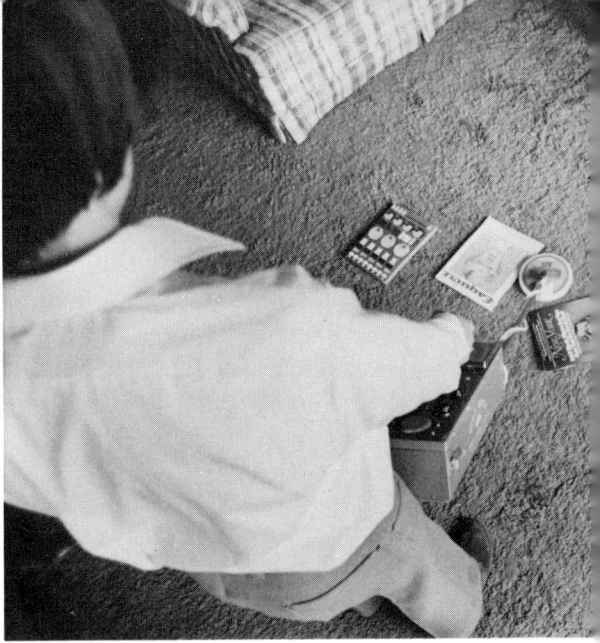

Experiment with the detector at home. *(George Sullivan)*

The BFO detector is different in that it *must* be operated with continuous sound. Adjust the tuning knob so that you get a fast beat, then sweep the coil over the concealed coin. Notice the increase in the frequency of the beats. It's not an abrupt change, however, but one that builds gradually and then dies.

Once you have the machine properly tuned, experiment with it. Scan the search coil over the living room carpeting, being careful to keep it an even distance from the floor. You should get an increase in the sound pitch (or in the frequency of beats) anytime you pass over nails in the flooring.

Don't make the mistake one new operator made. In testing his detector at home, he got a loud buzz from a particular spot at the base of one wall. In his imagination, he saw a cache of coins or jewelry concealed there by a previous owner. When he tore into the plaster, the only metal he discovered was a dustpan in a closet on the other side of the wall.

Or take the case of a New York apartment resident who kept complaining to the dealer who had sold him his detector that the machine was ultrasensitive, that whenever he tried it out at home it buzzed continuously. Yet at the dealer's showroom, the machine always worked perfectly. The two finally figured out that the detector owner lived in a building constructed of reinforced concrete, and the network of steel reinforcing rods within the concrete was causing the problem.

You can use the detector to trace the wiring in your home, beginning at an electrical outlet and following the conduits to the switchbox. You can locate the concealed pipes that make up your home's wiring system. And when you're going to hang a picture or mirror on the wall, you can use the machine to find out exactly where the metal studwork is located.

Many experienced treasure hunters become so sensitive to the signals given off by the metal detector, that they can actually tell scrap metal from coins, or one type of coin from another, all on the basis of the character of the sound they hear. You can work to develop this skill at home. Put several books on the floor; then turn your back and have someone put a dime under one book, a quarter under another, a bottle cap under a third, and etc. Then go to work with your detector to see how many items you can identify.

## TOOLS FOR DIGGING

Besides the detector the only other piece of equipment you really need is a digging tool. More than half of your time in any given searching foray is going to be spent digging, so it pays to get an efficient instrument.

If you're working in sand or very loose soil, an ordinary garden trowel may be all you need. Be sure that the handle feels perfectly smooth when you grip it. Any roughness is going to cause problems after an extended period of use.

With very loose sand, such as dry beach sand, it's a good idea to also use a sieve or straining device of some type. I've seen coinshooters use wire mesh deep-fry baskets as sifters. They bend the handle back on itself to reinforce it. Some equipment stores sell sifters for beach use.

In heavy soil, you need something sturdier than a garden trowel. The "All Pro Digging Tool" is recommended. One edge has been honed sharp so the instrument can be used in machetelike fashion in cutting roots or underbrush. It costs $7 and can be ordered by mail from Dumais Electronics, 37 Spring St., West Springfield, MA 01089.

You can also use an Army trenching shovel (each costs about $4) for heavy duty digging. Available in surplus stores, these shovels have a rugged steel blade that can be folded back onto the handle.

If you're looking for bottles, use a tool that's not going to cause any breakage, such as a hand pick. The one pictured, available at many equipment stores, sells for about $6. Coinshooting requires a number of specialized digging tools; these are discussed in Chapter 4.

It's not a good idea to keep your finds in your pockets. With all the bending and stooping you'll be doing, you're bound to lose things. Sling a drawstring pouch over one shoulder or fasten a small leather bag to your belt. Your local supply shop is likely to have such items in stock.

So much for equipment. The Do's and Don'ts of a typical mission follow:

DO know the machine. Read the instruction book carefully. Have the dealer explain the instrument to you. Operate it under his supervision, not merely in the store but outside it as well. Talk with other treasure hunters who use the same model detector; observe what they do. The

A sieve like this is what you need for dry beach sand. *(George Sullivan)*

A garden trowel will do if you're working in wet beach sand. *(George Sullivan)*

This is called the "All-Pro Digging Tool"; it's for heavy-duty work. *(George Sullivan)*

Leather pouch like this is ideal for what you find. *(George Sullivan)*

Hand pick aids in bottle digging. *(George Sullivan)*

Earphones assure peak performance. *(George Sullivan)*

more you know about the machine and its control system, the more success you're going to have.

   DO give the machine time to warm up. This is particularly true if you're going from indoors to outdoors, that is, from a relatively high temperature to a lower one. The detector's instrumentation needs time to adjust to the temperature variations. Stability of the signal can be adversely affected if you fail to wait.

   DON'T fail to test the machine before you begin searching. Bury several metal objects at various depths and seek to get a signal from each. One treasure seeker has a test "garden" in his backyard where he has buried several different items at measured depths. Before setting out on a field trip, he passes the search loop over the test plot to see whether the unit is functioning properly and the signal is what it should be.

   DO keep in mind that many different factors can affect the day-to-day operation of the detector. Weather conditions, for instance. On a wet or rainy day, you are likely to get a much stronger signal than when it's sunny and dry. Humidity, the makeup of the soil, and even altitude have an effect. With experience, you'll learn to anticipate signal variations.

   Once in the field, DO adjust the speaker control setting to what is called the threshold level, that point at which a signal is first heard. This assures maximum sensitivity. If the unit has a meter, it should read somewhere between zero and 10 (on a scale that reads as high as 50) when the machine is tuned to the threshold point.

   Once you've got the machine tuned properly, set the speaker volume control to the level you desire. Don't confuse the two. The first has to do with the sensitivity of the signal; the second with its loudness.

   DO consider using earphones. Not only will a headset enable you to hear the slightest variations in the signal and, thus, increase your finds, but they also aid you in concentrating. You become much more involved in what you're doing. And with earphones, you're also better able to tell when the signal starts to drift and the controls need adjusting.

   DO carry the detector in one hand, your probe or digging instrument in the other. Adjust the detector shaft to the proper length for your particular height. When you grasp the handle, you should be able to keep the search loop parallel to the ground and the appropriate distance from it without bending at the waist. When you keep erect, you're much less likely to tire.

   DO keep the search loop the same distance from the ground at all times. Failure to do this is common among beginners. You have to maintain the same distance; otherwise, you won't get the same signal all the time. And as this implies, you'll never learn to perceive the signal differences that result from objects of different kinds. You'll grow weary digging for metal pulltabs and small balls of tinfoil.

When using a large search coil—8 inches in diameter or bigger—keep the coil 3–4 inches from the ground as you sweep. This will help keep the coil free of mineralization interference and permit you to hear a deep response more clearly.

When using coils that are smaller than 8 inches, keep the search head flush to the ground, as long as there is no chance of damaging it. When you're working in a park or on someone's lawn, simply slide the coil along the grass. Never get more than an inch above the surface with a small coil.

When scanning level ground, DO walk straight ahead, moving the search head from side to side in wide sweeping arcs. Go as far to each side as you can. The distance between each sweep, each arc, should be equal in size to the loop you're using. With an 8-inch loop, keep the sweeps 8 inches apart (by making each of your steps 8 inches in length).

DO be deliberate with each sweep. When you pass the search head over a swath of turf and move it too fast, you can't help but miss detectable objects. And when it comes to coins, the ones you're missing are those that are buried deeply. And deeply buried coins are the ones that have been there the longest and have the greatest value.

At the slightest response, DO stop and dig. And once you've made your recovery, check the spot a second time. Something else may be there.

Before you begin digging, DO pinpoint exactly where the object is going to be found. Go back and forth over the area, tracing a series of zigzag patterns with the search loop. Dig in the exact spot where you get the strongest signal. In the case of a coin, the hole you have to dig shouldn't be much larger than the size of the coin itself.

DO be methodical. When you want to cover a large search area, first divide it into sections, marking each with stakes or stones, then scan each section carefully.

DO be persistent. One characteristic that distinguishes the experienced treasure hunter from the novice is tenacity. The veteran is relentless in covering an area which he has judged to be worthwhile, while the beginner is likely to be discouraged easily, switching abruptly from one spot to another, and never recovering anything of value as a result.

Even though other treasure hunters have worked an area, don't think that it won't continue to yield valuable items. A Hampton, New Hampshire, treasure hunter declares: "A man who lived near a park where I do my hunting once told me that I wouldn't find anything because he often saw people with detectors there. Well, I've been there

Store detector like this. *(George Sullivan)*

nine or ten times myself and I never fail to find something. I've taken home more than 400 coins, including about 40 Indian Head cents."

DON'T use the machine if the batteries are weak. You'll just be wasting your time. One dealer says this: "More than 90 percent of my customers' service problems have to do with weak batteries or dead batteries." If you have the slightest doubt about the amount of energy remaining in a battery, throw it away. If you plan to be in the field for an extended time, be sure to carry spare batteries.

When your detector is not being used, DO stand it on the control box in an upright position.

When you plan to store the detector for more than three or four weeks. Do be certain to remove the batteries from the battery compartment. Store them in a cool, dry place.

DON'T expect to become a millionaire.

# Chapter 3

# WHERE TO SEARCH

One of the most unusual treasure finds recorded in recent years occurred in 1972. The finder was Francis Sherman of Eugene, Oregon. Sherman had heard about treasure hunters finding pots, kettles, and other objects left behind by the historic Lost Wagon Train that had struggled to cross the Cascade Range through the Willamette Pass in the winter of 1853. After research, he and some friends decided to probe a section of the trail high in the mountains near the town of Oakridge.

Sherman was scanning the moss-covered ground near an ancient tree stump when suddenly his detector started humming. He and his companions quickly peeled back the moss and started digging into the soft black dirt beneath. About 16 inches down, they came upon a cache of 14 "coins," some made of silver, some made of gold. But they were not a type of coin that is listed in any catalog. Each was about the diameter of a half dollar but two or three times as thick. Sherman thought at first that they were ingots of some type.

One of the pieces of metal was inscribed with the initials J.S. and the date 1851. Sherman later learned that one of the riders of the Lost Wagon Train was named John Stewart. When fierce winter storms struck, the party began abandoning possessions to lighten their load. Many people died but a handful survived, thanks to the efforts of a rescue party from the Willamette Valley. Sherman believes that John Stewart buried the coins with the hope of returning some day and recovering them. His research suggests that these curious pieces of gold and silver may have been used for transactions among wagon train parties.

Newspaper accounts of Sherman's find spoke of his good fortune. If luck was involved, it was only to a small degree. Before they set out, Sherman and his friends spent weeks in investigation and inquiry. They didn't know the exact spot to look, of course, but they had a good idea of the route taken by the members of the ill-fated party, and they followed it step for step.

Or take treasure hunter Gerald Rainey, from Wallingford, Connecticut. Not long ago he obtained old maps of the Wallingford area. They showed roads that no longer existed, the sites of old schoolhouses, and the former location of Wallingford's railway station. Around the foundation of what used to be a workshop in the 1800s, Rainey discovered several ax heads under a foot of soil. They're worth $40 each. He also found nineteenth-century coins and old cutlery.

If Rainey has any "secret," it's the work that he does before he heads out into the field. He says that he spends 90 percent of his time researching.

Books, literally hundreds of them, magazine articles, treasure-hunting magazines, treasure-hunting clubs, and several departments of the federal government—any and all of these can provide you with information and what to look for and where to look for it.

## BOOKS AND LIBRARIES

Well over a thousand books have been written on treasure hunting and topics related to it. Some concern lost mines and outlaw caches, others have to do with gold panning and placer mining. Coinshooting is becoming a popular topic; sunken treasure has always been one. Many of these books provide clues to where treasure can be found. A sampling of available books is to be found in the appendix of this book.

Where do you obtain books about treasure hunting? Many are mail-order titles, available from firms that specialize in the sale of treasure-hunting equipment and supplies. The widest selection of books is available from these three firms:

Treasure Book Headquarters, P.O. Box 328, Conroe, TX 77301;

Keene Engineering, 11483 Vanowen St., North Hollywood, CA 91605;

Eight States Associates, P.O. Box 1438, Boulder, CO 80302.

In each case, write and request a free catalog.

Don't fail to check the books at your local general reference library or the reference department of your branch library. Their card catalog is an alphabetical author, title, and subject index of all the books in the library's collection. Look under the heading "Treasure Trove."

The books in your local library represent only a small sampling of what's available to you. Some libraries have printed catalogs that list the holdings of other libraries. The most notable catalog of this type is the

Treasure hunting in Oregon's back country. *(White's Electronics)*

U.S. Library of Congress *National Union Catalog.* It lists the books in the huge collection of the Library of Congress and other cooperating libraries.

Once you've established the name and location of the library that has the book you're seeking, you can probably obtain it through an interlibrary loan. Most libraries lend material to other libraries for use by qualified individuals. Usually you can retain such borrowed material for periods of up to four weeks. Inquire at your local library if this service is available.

Be sure to find out whether there is an historical society in your area. Many of these organizations offer the use of historical material which can be of extreme value, no matter what type of treasure you're seeking. Inquire at your local library about local historical societies or ask to consult the handbook, *Directory of Historical Societies and Agencies in the United States and Canada.* Published by the American Association for State and Local History, the guide lists over five thousand historical associations, describing each and the services it offers.

Some historical societies are only for the use of individuals doing scholarly research, they do not welcome treasure seekers; others serve as depositories for official state records and, thus, what they have available is of quite a limited nature. However, institutions in these categories are in the minority.

"We have excellent collections of printed materials, manuscripts, maps, etc., relating to Indiana's past," says Herbert Hawkins, Executive Secretary, Indiana Historical Society (State Library and Historical Building, Indianapolis). "These resources are available to all responsible researchers and could definitely be helpful to treasure seekers."

Mr. Hawkins adds: "The prospective seeker would need to have some idea as to what he is looking for. If someone came in and announced that he had just bought a metal detector and wanted a list of treasure sites, I'm afraid we would not be very much help."

The Florida Historical Society (The University of South Florida Library, Tampa) is also ready to serve. "The resources of our collection are available for use by any interested person, including treasure hunters and amateur archaeologists," says Paul Eugene Camp, Assistant Special Collections Librarian. "Our materials are used quite regularly by people seeking old towns and fort sites, among other things."

Among the wide range of materials available, Mr. Camp cites books relating to treasure hunting in Florida's old towns, old forts, and etc. "We also have a fair-size collection of Florida maps, the earliest dating from 1584. Additionally, we have complete sets of the U.S.G.S. quadrangle maps of Florida and the C.&G.S. nautical charts of Florida waters." (See more details on this immediately following.)

Keep in mind that city, county, and state records are public records, and you have the right to consult them. Often, too, you have the right to check the records of the county sheriff or county attorney. Of course, any person is able to consult official community maps containing building locations and property lines.

Newspaper files that libraries maintain can be helpful, too. Or perhaps you can gain access to a newspaper's reference files, called the morgue. The newspaper librarian can tell you what local policy is. The file is likely to consist of hundreds of envelopes arranged by subject and containing clippings from the newspaper and other sources. The news stories you'll want to read will be contained in envelopes titled, "Buried Treasure," "Treasure Finds," or, simply, "Treasure."

For news stories concerning important treasure finds of the past anywhere in the United States, you can consult the *New York Times*. Your library undoubtedly has a microfilm edition of the newspaper. The *Times* is unique because it is the only American daily newspaper with a printed index. By consulting it, you can obtain the dates and circumstances of virtually any important treasure find. Look in the index under the heading "Treasure, Hidden" or "Treasure, Sunken."

Of course, neither your local newspaper or past issues of the *New York Times* can inform you of the location of any secret treasure caches. But they can be a starting point.

## MAGAZINE RESEARCH

Articles that have appeared in American magazines are another excellent starting point. The standard guide to such articles is the *Readers' Guide to Periodical Literature*. It indexes approximately 130 publications by subject and author from the year 1900 on.

Articles about treasure hunting are listed under the heading "Treasure Trove." This is a sampling of listings from the *Readers' Guide* for the period from March 1963 to February 1965:

"Anyone for Buried Treasure?" (Cocos Island), *Saturday Evening Post*, June 13, 1964.
"Bonanza on the Bottom," (Scuba divers on Florida beaches), *Time*, Aug. 25, 1964.
"Deep Sea Diver Uncovers Treasure," (Sixteenth-century Spanish ship on Bermuda reef), *Hobbies*, November 1964.
"Drowned Galleons Yield Spanish Gold," *National Geographic Magazine*, January 1965.
"King John's Lost Treasure," *Saturday Evening Post*, May 30, 1964.
"Oak Island's Mysterious Money Pit," *Reader's Digest*, January 1965.
"Spanish Gold Two Fathoms Deep off Florida Coast," *Saturday Evening Post*, December 12, 1964.
"Treasure Fever Hits Florida Coastline," *Business Week*, October 10, 1964.

Let's suppose you dig out the names and dates of several magazine articles that you think will be helpful to you, but your local library doesn't have the magazines. Consult the *Union List of Serials*, a catalog that gives the location and inclusive dates of the holdings of periodicals in American and Canadian libraries. Using this reference work, you may be able to locate the periodicals you need in a nearby library.

If the library where the periodicals are located is some distance from you, write and inquire whether you can obtain a Xerox copy of the material. Some libraries, particularly those in big cities, provide this service. They base their charges on the number of pages ordered, so if a good deal of material is involved you may want an estimate first.

Also be aware of the *Nineteenth Century Readers' Guide to Periodical Literature*, an index of periodicals from the years 1890–1899. Articles are indexed by subject and author.

A third periodical index that can be helpful is *Poole's Index to Periodical Literature*. This is an index to American and British periodicals for the period 1802 to 1906. It is the standard source for magazine articles of the nineteenth century. A subject index, the articles in which you'll be interested are again listed under "Treasure."

If your treasure hunting is going to involve Canada, you'll want to consult the *Canadian Periodical Index, 1928–1947*, and the *Canadian Index to Periodicals and Documentary Films, 1948-present*.

## TREASURE-HUNTING MAGAZINES

Most of the magazines devoted to treasure hunting concern such topics as ghost towns, lost mines, and such. But two publications, *Treasure* and *Treasure Search*, cover the field with a good amount of thoroughness and provide much helpful information and advice. A complete list of treasure-hunting magazines appears in the appendix.

White's Electronics (1011 Pleasant Valley Rd., Sweet Home, OR 97386) publish a twelve-page bimonthly newsletter, *Discover*. It's filled with news about unusual finds made by the company's customers, as well as background information on coins, buttons, meteorites, military artifacts, and etc. The newsletter is free; write and request that your name be put on the mailing list.

"The Treasure Index of Current Finds," a service which began operation in 1974, provides subscribers with information on where recent treasure discoveries have been made. Locations of these finds are noted on state maps. When the reported finds become centered in one area, it is said to indicate where other discoveries can be made, or, as a promotion brochure for the service states, "The secret rests with the clusters...[a] clear indication of fertile territory." For more information, including subscription rates, write Treasure Index of Current Finds, P.O. Box 101, Bronx, NY 10468.

## CLUBS

If there is a treasure-hunting club in your city, consider yourself fortunate. Its members are sure to be familiar with most of the search sites in the area, and while no one is going to reveal secrets to you, you will be able to obtain a good deal of information from attending club meetings ...information on where to search and on treasure-hunting equipment and its use.

You should also be aware of two national organizations. One is the National Treasure Hunters League (P.O. Box 53, Mesquite, TX 75149), a club that publishes a quarterly magazine for its members that features articles on treasure-seeking forays in different parts of the United States. Membership is $5 a year.

The NTHL, through its "regional offices"—dealers, actually—also sells detectors, not one brand or two but those manufactuted by almost all major detector firms. When you purchase a detector through the NTHL, you receive a free membership in the organization.

Finally, the NTHL offers an information service. Members are encouraged to write or call Ray Smith, the league's head man, to obtain answers to their treasure-hunting queries.

The Prospectors Club International (P.O. Box 68069, Indianapolis, IN 46205) has as its intent the exchange of information among members. This is accomplished not through meetings or an annual convention but by means of an eight-page quarterly newsletter, *The Prospector.* A one-year membership, which cost $4 in 1974, also gave you the right to query the club president, on any aspect of treasure hunting from equipment required to possible locations.

You're likely to derive the most benefit, however, from a local or state club. A list of clubs is to be found in the appendix.

## MAPS

For anyone treasure-hunting in a remote section of the country, the best maps are the remarkable topographical maps prepared by the U.S. Geological Survey. They're remarkable because of their elaborate detail. They show roads, bridges, streams, swampland, fence lines, mineshafts, and even houses and barns. Nothing else compares with them.

These maps are prepared in six scales, from 1:24,000 (meaning that every unit of measurement on the map is represented by 24,000

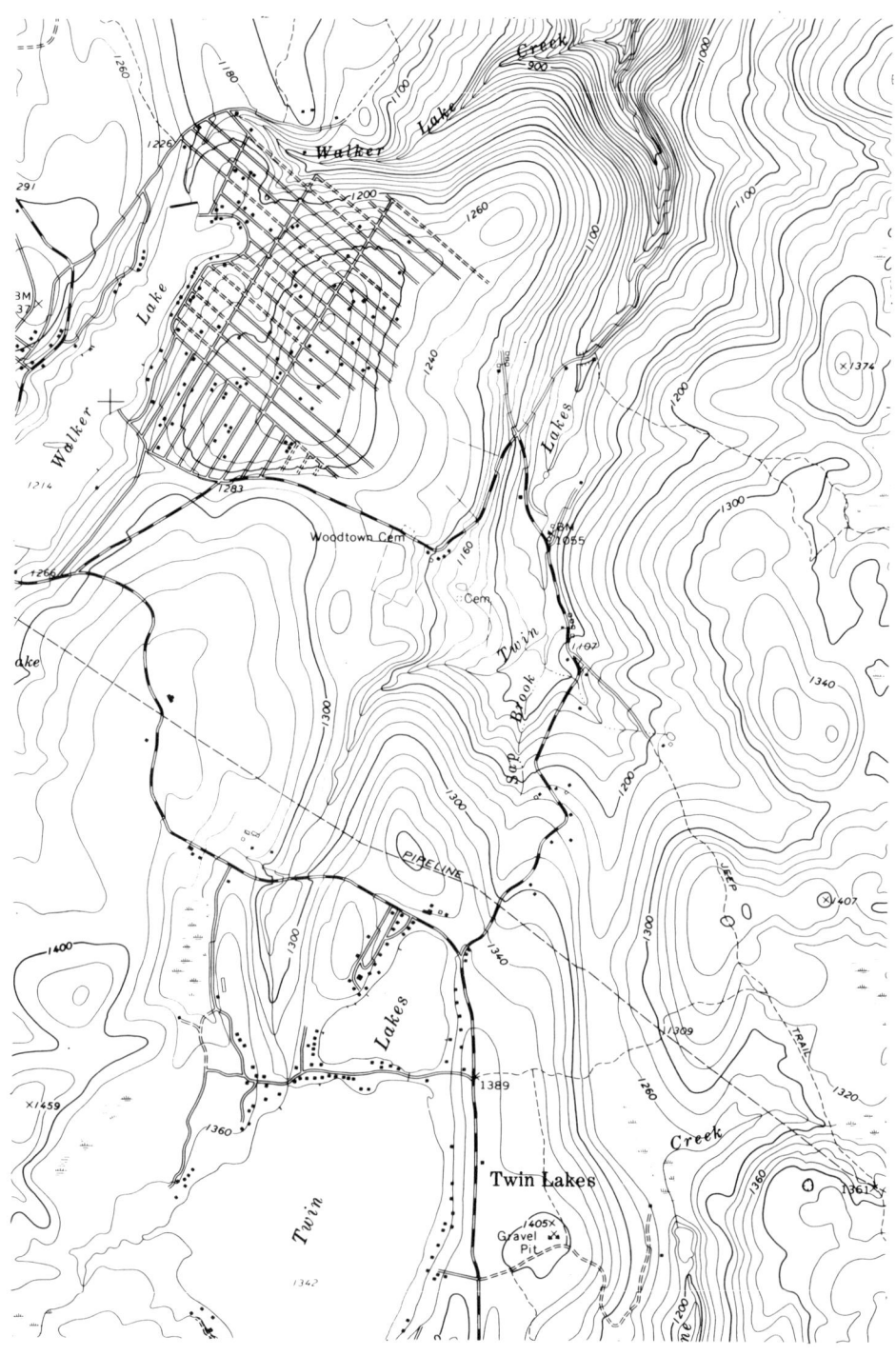

U.S.G.S. maps are prepared in elaborate detail. *(U.S. Geological Survey)*

corresponding units on the earth's surface) to 1:500,000. With the former, an inch equals about ⅜ of a mile (2,000 feet); with the latter, an inch equals about 8 miles (42,240 feet).

A map with a scale of 1:24,000 for the state of Delaware would be so large that you'd have to have a couple of football fields to lay it all out. So for the sake of convenience, the Geological Survey issues these maps in sections called quadrangles. Each quadrangle costs 60¢ to 75¢.

To secure these maps, write to the U.S. Geological Survey (GSA Building, Washington, DC 20242) and request an index to the topographical maps of the state or area in which you're interested. The index will enable you to order the quadrangles you need.

There are several regional U.S.G.S. map distribution centers, and any one of them will supply you with an index and the particular maps you need. The main distribution office for maps covering areas east of the Mississippi River is: Distribution Section, U.S.G.S., Washington, DC 20242. For maps covering areas west of the Mississippi, the address is: Distribution Section, U.S.G.S., Federal Center, Denver, Colorado 80225. For Alaska, the address is: U.S.G.S., 520 Illinois St., Fairbanks, Alaska. There are also map distribution centers in Dallas, Texas; Salt Lake City, Utah; Spokane, Washington; Menlo Park and San Francisco, California; and Anchorage, Juneau, and Palmer, Alaska.

Large map retailers—listed in the Yellow Pages under "Maps"—sometimes have such maps for sale. They are also sold at most National Forest and National Park headquarters (listed in your telephone directory under "United States Government, Interior Department").

Canadian maps comparable to U.S.G.S. maps are available through the Map Distribution Office of the Department of Mines and Technical Surveys (615 Booth St., Ottawa, Ontario). In your letter requesting information, make reference to maps of the National Topographical Series.

By studying U.S.G.S. maps, you can form a precise idea of the terrain. This is because these maps include carefully drawn contour lines, lines which connect points of equal altitude. The lines are broken at intervals by figures indicating the height in feet above mean sea level.

These contour lines are unusual because there are so many of them; they are drawn at intervals representing distances of ten to eighty feet. You can tell at a glance where elevations are located and, in each case, how steep the climb is. Study a map carefully, and when you're out on the trail no detail of the terrain should surprise you.

While cartographers at the Geological Survey work continually to keep maps up-to-date, many of the quadrangles now being sold were prepared twenty or twenty-five years ago. This fact can work to your advantage, however. One treasure seeker tells of using a quadrangle that covered a section of Hampden County in south central Massachusetts, just outside the town of Monson, which had been prepared in 1949. The map gave a location for a farmhouse that was no longer there, but inspection of the site disclosed a building foundation and enough charred remains to indicate that the structure had been destroyed by fire. In going over the area with a detector, two cast-iron spiders were found. Spiders are frying pans with long handles and short legs that are often prized as antiques. So U.S.G.S. maps are not only valuable for their detailed analysis of the terrain of any given region, they also furnish clues on where to search.

Many other branches of the government publish maps which are highly regarded for their elaborate detail and accuracy. The U.S. Forest Service (Washington, DC 20250), local ranger stations, and regional field offices of the Forest Service (listed in Chapter 6) sell and distribute maps. Maps can also be obtained from the Department of Defense (Washington, DC 20301), Department of Commerce (20230), and the Post Office Department (20260). Some of the maps available from these sources depict old trails, roads, railway lines, schoolhouses, and churches. Frequently, however, only foundations of these structures remain.

Don't overlook your local library as a source of helpful maps. You're not likely to find treasure maps with a bold X marking the spot to dig for fabled riches. Maybe what you'll be shown is an old map of your state dating, say, to the nineteenth century. Look it over carefully. It may indicate many structures that have vanished: railroad depots, sawmills, telegraph offices, and etc. Any one of these sites could provide excellent hunting.

### ABANDONED TOWNS AND BUILDINGS

The dictionary defines a ghost town as "a boom town of the old West that has been completely abandoned." To most treasure seekers, however, the term has a much broader meaning.

For one thing, ghost towns are not always *completely* abandoned. Sometimes a resident or two or a family occupies a central building. There may even be a sufficient number of people to sustain a post office, a store, and perhaps an eating place. Some of these towns have been "restored" by local residents and seek income as tourist attractions. Virginia City, Nevada, is in this class. In the heart of the fabled Comstock Lode region, Virginia City was perhaps the richest of all the mining towns of the mid-1800s. Now it thrives on money spent by visitors from every part of the country.

But many towns of the West have been, indeed, totally forsaken, and the old houses, stores, and other structures stand in ruin and decay. What caused these towns to fail? Something happened to halt the flow of money into the town's economy. A nearby mine ran out of ore. The forests were cut down. A railway or highway passed the town by.

When you come upon a ghost town, it is likely to be in sad condition, bearing evidence that it has been frequently subjected to the ravages of looters. The floors of buildings may be torn up; the walls caved in.

This abandoned mill site in Massachusetts yielded a score of nineteenth-century artifacts. *(George Sullivan)*

Adobe walls of former jailhouse in the ghost city of Hillsboro, New Mexico. *(State of New Mexico, Department of Development)*

Sometimes the destruction is so senseless that it strains belief. A husband and wife treasure-hunting team, vacationing in Arizona, set out to find Swansea, an abandoned mining town that they had read about, located in the northwest corner of the state. With the help of residents in towns nearby, they finally reached their destination. What they found shocked and saddened them. Vandals had burned the framework of adobe buildings. They had dug up the graves in the town cemetery and destroyed monuments. Even the town dump didn't escape; bottles and jars—every glass object—had been wantonly destroyed. Unfortunately, this type of vandalism is not uncommon. The Bureau of Land Management tells of treasure hunters who have used bulldozers to excavate Indian ruins and old buildings.

There is no "typical" ghost town. Those in Washington and Oregon are very different from those in the Southwest. The town of La Paz, Arizona, located on the Colorado River about ten miles north of Ehrenburg, was once the most important city in the state, now only adobe ruins remain. The structures that still stand in Rich Hill and Bradshaw City, both in the central part of Arizona near the once opulent Crown King Mine, are of adobe, too.

California's ghost towns have more of a classical appearance. The town of Halivah, about 25 miles east of Bakersfield, once well known for its high-grade ore, boasts many old wooden buildings that are completely intact. About 40 people live there. Wood structures were also common to Nevada ghost towns. But sometimes only walls and crumbling foundations remain. Such is the case in Oriental, Nevada, which is near the California border about midway between Las Vegas and Reno. It's mostly debris.

There's nothing secret about the locations of the ghost towns of the West. Many are so well known they are listed on service station road maps. Some western states provide free literature on their ghost towns. Arizona, for example, offers a booklet titled "Ghost Towns and Lost Treasure." It's available from the state's Department of Economic Planning and Development (3003 North Central Ave., Phoenix, AZ 85015). Colorado's Division of Commerce and Development (602 State Capitol Annex, Denver, CO 80203) lists the state ghost towns in its brochure, "Colorful Colorado Invites You."

Another way to learn more about ghost towns is through a series of books available from White's Electronics (1011 Pleasant Valley Rd., Sweet Home, OR 97386). There are individual books for these states: Arizona; California; Colorado-Utah; Montana-Idaho-Wyoming; Oregon; Nevada; New Mexico-Texas; and Washington. Except for the book on California which costs $3.95, they're priced at $2.95 apiece.

Abandoned buildings are everywhere, not just in the Old West. Chet Blanchard, a detector dealer in Minerva, Ohio, used one of his machines to survey the foundation of an old shack near his home that had been vacant for fifty years. When he passed the search loop over the front steps of the shack, he got a loud signal. Beneath the steps he found a tractor toolbox. Inside the box was a rolled up copy of *Lone Scout* magazine dated 1919. Rolled inside the magazine was a Mason jar filled with old coins—33 silver dollars, almost all of them dated before 1900, 55 Indian Head cents, and several commemorative tokens, including a Grover Cleveland presidential token dated 1884.

When searching an abandoned building, you'll find that either a TR or BFO detector functions with about equal efficiency. An 8-inch search coil is likely to suit your purposes best. It's small enough to respond to buried coins but big enough to detect larger objects at a reasonable depth. Use a unit with a swivel head. This enables you to keep the search loop level when probing hard-to-get-at places such as ceiling beams and foundation piers.

You needn't use a machine that's equipped with a discriminator, that is, a detector that rejects "junk" items. In fact, it can be detrimental. You're seeking metal objects of all types; you don't want to overlook anything.

Do a thorough job. Check the fireplace. Check the area about the hearth with particular care, and pass the detector up inside the flue as far as it will go.

Examine the beams. Look on top of the beams. Sometimes valuable items were stored on the beams... and left there. Look closely at each beam to see whether there are any filled-in holes, a sign that something may be hidden within.

Use the detector to scan the windows and doors. Look carefully where the sash weights might be located. Newspaper stories recently hailed the searcher who recovered a pair of weights which were pure silver painted black. Small caches of coins were sometimes hidden under loosened window boards or at the top of door casings.

Staircases should be checked. The bottom step was a favorite hiding place. Treasure seekers have also told of finding coins and jewelry in hollowed out stair railings.

Don't fail to scan the building's foundation. Jars of coins were sometimes placed behind loose stones for safekeeping. Look under the porch, in any outdoor fire pit, and also around fence posts and gateposts.

Whenever you're searching an abandoned townsite or homestead, try to find out where the people dumped their refuse. Dumps that are fifty or so years old, often give up objects that are now prized by collectors and antique dealers (see Chapter 5).

The whole business of putting one's rubbish out on the street and having someone haul it away is a relatively recent concept, particularly in most suburban areas of the country and especially in most rural ones. It used to be that the individual householder had to get rid of the family refuse himself.

What did people of two or three generations ago use for dumping areas? One common method was to dig a pit at some distance from one's homesite; the pit was used as a dumping site until it was filled, whereupon it would be covered with soil and another pit dug. Sometimes these trash pits can be recognized by depressions left in the ground as a result of the settling of the rubbish. Don't fail to scan such a depression carefully with the search loop.

Sometimes these pits were dug as deep as seven or eight feet. If there happened to be a natural depression in the earth, or when there was a ditch, rock crevice, or ravine nearby, the householder would frequently use these as dumping grounds. Pieces of glass or chips of porcelain sometimes offer a clue as to where these areas existed.

Some people took pains to camouflage dumping areas. They would dump trash into stone walls and cover it with a layer of rocks. They would haul it into a thickly wooded area and dump it there. Look for small secondary paths that lead off of well-traveled trails as a tip-off to these sites.

Whenever you come upon a former dumping ground, dig into it carefully. Carry along a pair of work gloves so that you can hand dig. With a shovel, you're likely to damage anything that's metal and run the risk of breaking valuable bottles or glassware.

While ours is sometimes referred to as a cashless society, with financial transactions carried out by check or credit card, such wasn't always the case. Not too many decades ago, the number of banks was much smaller than it is today, and those that did operate were not always held in high esteem. Only a few people had checking accounts.

This meant that families had to keep sufficient cash on hand for day-to-day expenses. The head of the family would often seek a hiding place for these funds, usually one that was fairly accessible. In rural areas, it was common for a man to take a sturdy container, such as a heavy cooking pot or an iron skillet, place the money inside and bury it. He didn't go very deep, only one or two feet. After all, he had to be able to get to it readily. When a farmer needed seed, say, he didn't want to use valuable time digging for the money to pay for it, nor did he want to risk the chance of being seen. He'd slip out of his house under the cover of darkness, remove the sum he needed, and carefully cover the "strongbox" again.

The posthole bank was another favorite hiding place. A farmer or homeowner would put his money or valuables in a jar or other container and then "deposit" them beneath a fence post, first digging the posthole a little deeper to allow for the container. To distinguish the post from the others, he would sometimes nail a horseshoe to it, or he might notch it with an ax. When he wanted to make a "withdrawal," he sometimes had

Check all overhead beams carefully.
*(George Sullivan)*

Window casings should be scanned, too.
*(George Sullivan)*

Don't overlook building foundations.
*(George Sullivan)*

to cut the barbed wire in order to be able to remove the post from the hole, so spliced wire sometimes tips off where the bank used to be—or is. Slack wire is another clue.

Watertight containers filled with money and valuables were sometimes dropped into cisterns or wells. Windmills were sometimes used as hiding places, and hardly a year goes by without one reading about a find being made in or around a tree stump. Near DeLand, Florida, a vacationer from Pennsylvania was demonstrating a new metal detector for a friend along a tree-lined path when the unit suddenly responded to a hoard of jewelry which had been stolen several years before. The thieves had used a hollow tree as their hiding place.

Flower beds and former vegetable gardens are always worth searching. A person hiding valuables could dig in these places without ever arousing suspicion.

The sources mentioned in the opening pages of this chapter are sure to provide you with information as to where to find towns and homesites of the past. Or you can simply follow the lead of a Bronx, New York, photographer who does much of his treasure hunting in the Catskill Mountains. He likes to search the foundations and grounds of abandoned buildings and he has a simple method of finding them. "I just get into my car and drive over backroads until I come upon a good site," he says. "It might be easier if I used old maps and planned in advance where to go. But my method has never failed yet."

### TREASURE HUNTING AND THE LAW

One concept is basic to all treasure hunting: All land in the United States is owned, owned by an individual, a company or a corporation, or by a local, state or the federal government. This doesn't necessarily work to your disadvantage. It does imply, however, that you must be aware of who owns the land you plan to search and, when required, you obtain the approvals and permits that may be necessary.

This applies almost universally in the case of private land. Get permission in advance. When you approach the owner, be completely honest about what you're searching for. Usually he'll give permission with a shrug or a nod of his head.

When the object of the search is something very worthwhile, a treasure cache of some type, let's say, it's wise to have a written agreement with the landowner. Many treasure-hunting clubs have agreement forms available or they can be purchased at equipment shops. Such agreements set down the conditions of the search, usually specifying that the searcher will leave the property in precisely the condition he found it, and the property owner is not liable in any way for injuries that might arise out of the search. The agreement also specifies how the treasure, if recovered, is to be divided.

The statutes that apply to treasure hunting on state-owned land are usually contained in the antiquities code of the state. More and more states are adopting such laws. The Texas Antiquities Code, which dates to 1969, is one of the most restrictive. It designates as "state archaeological landmarks" not only all state-owned land, but also all land belonging to

Use a long probe like this one when you think you've come upon a bottle dump. *(Garrett Electronics)*

Stone walls were often used as hiding places. *(George Sullivan)*

any county, city, or political subdivision of the state. The state's Antiquities Committee serves as the legal custodian of any items recovered, not merely buried treasure, but also "items relating to the Spanish exploration of Texas, historic battlegrounds and fort sites, pretwentieth century shipwrecks, and American Indian dwellings." If you want to do anything more than coinshoot in Texas, you'd better get permission from the Texas Antiquities Committee (P.O. Box 12276, Austin, Texas 78711). Violations of the statute are punishable by "a fine of $50 to $1,000 and/or a jail sentence of up to 30 days."

New Jersey's statutes concerning recovered valuables go back to English common law: the surface owner has the rights to anything that lies below the surface. The Proprietors of East Jersey and the Proprietors of West Jersey maintain deed books which date to the founding of the colony. "Amazingly enough," says a spokesman for the state's Bureau of Geology and Topography, "they sometimes are still able to prove a claim to title."

In the case of coastal states, the antiquities statutes invariably apply to shipwreck areas, imposing penalties for the unauthorized removal of artifacts. Violating such statutes in Florida can lead to a $5,000 fine, "plus seizure of all boats, equipment, and instruments used in connection with the violation."

The answer to all of this is to be aware of state legislation. If you're not sure of what the law says, write to the attorney general of the state and inquire. Treasure-hunting clubs are usually well informed on local laws. For general information on the topic, obtain a copy of the *Treasure Hunter's Yearbook* (P.O. Box 1215, Odessa, Texas 79760). It contains a state-by-state rundown of existing legislation. The book costs $4.

When it comes to federally owned land, most areas are open to treasure hunters, except national parks. Here the regulations are particularly stringent. Park officials simply don't want any treasure hunters. Period. There are even regulations covering the mere *possession* of metal detectors within national park boundaries.

In the case of National Forest lands, the situation is different. Treasure hunters are generally welcome or, at least, not unwelcome. But it is always a good idea to check in advance with the forest supervisor or ranger, or with the appropriate regional office (see Chapter 6). The same is true of the lands under the jurisdiction of the Bureau of Land Management. In general, there are no restrictions, but contact the regional office before you begin your search.

If you plan to search on federally controlled land, you come under the jurisdiction of the American Antiquity Act of 1906, which states in part:

"Any person who shall appropriate, excavate, injure or destroy any historic or prehistoric ruin or monument, or any object of antiquity, situated on land owned or controlled by the Government of the United States, without the permission of the secretary of the department of the government having jurisdiction over the lands on which said antiquities are situated, shall, upon conviction, be fined a sum of not more than $500 or be imprisoned for a period of not more than 90 days, or shall suffer both fine or imprisonment at the discretion of the court."

The permits referred to here are issued only to recognized scientific and educational institutions, not to individuals. Thus, you can be caught in a situation where you are permitted to use a metal detector but can be prosecuted for the removal of discovered artifacts. The only solution is to establish in advance what statutes apply.

It is a matter of fact that federal, state, and local legislators are becoming increasingly wary of treasure hunters, and more and more restrictive legislation is what is resulting from this apprehension. The fears of government officials are not without justification. Much wanton destruction has been wrought in the name of treasure hunting, especially in the West. "Treasure hunting in Nevada has done more damage to ghost towns than any other single activity," says Donald R. Tuohy, Curator of Anthropology for the Nevada State Museum. "...anyone advocating treasure hunting in Nevada is encouraging the destruction of our historic past."

What you must do to operate within the law. This means obtaining permission before entering or digging upon private property, and adhering to the various local, state, or federal statutes.

# Chapter 4

# ALL ABOUT COINS AND HOW TO FIND THEM

Of the thousands upon thousands of people who now list coinshooting as their principal avocation, Earl Clopine of DeLand, Florida, is fairly typical. He concentrates his efforts on Daytona Beach's twenty-three miles of coastline.

Earl bought his first metal detector just before his retirement as a warehouse manager several years ago, but within a few months he purchased a more efficient machine that produced far better results. He now describes that first machine as "over-priced junk." He has since traded upward several times and now owns two deluxe models, each of which cost about $200.

Earl's first find was a Jefferson nickel. So was his second. Small change constitutes the great bulk of what he finds. On a poor day he will turn up 30 to 40 coins; on a good day, 150 to 200. Of course, his finds have paid for his equipment many times over.

His most valuable find was a Swiss-made wristwatch worth several hundred dollars. Fortunately, it had not been touched by sea water, which probably would have rendered it worthless, or just about so. As far as coins are concerned, his oldest discoveries have been an 1853 dime and a 1921 half dollar.

Most of the jewelry he has recovered has been of the dime-store variety. However, he's been able to sell many of the hundreds of rings he's found for their gold content. These are usually wedding, school, or college rings that have apparently slipped from bathers' wet fingers and eventually washed onto the beach. He has found as many as four rings in a single search area.

While he prefers the beach, he occasionally transfers operations to a local park or picnic area, to church grounds or school grounds. It's on these sites that he has found his oldest coins, many dating to early in the century.

To avoid these, pick virgin territory, if you can. *(George Sullivan)*

In the beginning, Earl was plagued by signals from soft-drink and beer-can pulltabs and by the cans themselves, but he has learned to distinguish between signals of different types and he knows when to dig and when not to. He has always used a TR machine. Does he employ constant sound and use earphones? Of course he does.

Earl says he gets his best results with a search loop that penetrates six to ten inches. He's gone deeper by using bigger loops but these make it difficult to pinpoint his finds. In addition, at greater depths he gets more signals from things like rusted bedsprings and old oil drums.

Like most coinshooters, Earl says that the real rewards he gets from the hobby are other than monetary. His detector keeps him out in the sunshine and provides him with a healthy amount of exercise. And unlike some Florida retirees, Earl never has a problem trying to decide what to do with his leisure time.

For coinshooting, beaches are more popular than any other site. You can hunt virtually all year round. *(George Sullivan)*

Earl has been successful as a coinshooter because he owns quality equipment and is skilled in using it, and also because he is realistic with himself about what he expects to find. A down-to-earth attitude helps to develop diligence and patience.

Whether or not you are as skilled and experienced as Earl Clopine, the penny is the coin you are going to find most frequently. The chief reason for this is the lower the value of the coin, the more are minted. Thus, more are available to lose. But that's not the only reason. When a person drops a half dollar, say, it doesn't get hidden in the grass or buried in the beach sand the way a dime or a penny does. And even if the larger coin does happen to get lost, it's usually worth the person's time to search for it.

What might be described as "average" results were achieved by a 41-year-old Shelby, Ohio, coinshooter. Using a Goldmaster detector, and working an hour or two once a week over the period of a year, he recovered 694 coins: 558 cents, 53 nickels, including 3 Liberty head nickels, the oldest dating to 1899, 58 dimes, 19 quarters, and 6 half dollars, all of them silver. (Of course, he's found countless other items, including several rings.)

A very active Columbus, Ohio, treasure hunter kept a careful tabulation of all the coins he found over a period of several years, and developed a statistical breakdown. For every hundred coins he recovered, he found 74 cents, 11 dimes, 10 nickels, 4 quarters, and 1 half dollar.

Not only will you find more pennies than any other coin because so many more are in circulation, there's a scientific reason. Copper is of the least stable metals of which our various coins are minted. If a cent happens to drop in wet soil, corrosion begins almost immediately. The wetter the soil the more the corrosion. In time, the soil surrounding the cent becomes saturated with copper salts and thus the search loop is able to perceive the coin more easily. A cent that has been buried for a long time period in damp ground gives off a signal that is as strong as that of a half dollar's at the same depth.

Gold coins are not subject to corrosion. Though buried for years, a gold coin can remain bright and shining and the engraving may stand out boldly. Coins lost on a beach—gold, silver, copper, anything—become badly worn, however, because of the abrasive action of sand and surf.

Silver coins retain their brilliance if buried in a dry place, but soil that is damp and high in mineral content will tarnish and even corrode them. A coinshooter who searches the desert wastelands of southern California reports that the soil there is so highly mineralized that it "eats up" pennies, leaving only a granular green disc in the soil.

When working in offshore waters, be sure the search coil is waterproof, not merely water resistant. *(Search Electronics)*

Parks provide a great deal of potential, too. *(George Sullivan)*

Use a search loop that is going to provide you with at least eight inches of depth. (The matter of search coils is covered in Chapter 1). The newer coins, generally speaking, are closer to the surface. The older, more valuable ones, are deeper down.

Cold weather has little or no effect upon coinshooting. But when the temperature slides to below 32°F, it's a different story. The ground freezes and you can't dig, or at least it may be very difficult to do so.

Wet weather is no problem. In fact, it helps. When the ground is damp, it transmits detector signals with greater than normal facility. You're able to spot coins that are buried deeper, and those at shallow depths give off stronger signals than normal. Most coinshooters welcome rainy weather.

If you plan to work when it's wet and your search loop isn't waterproof, simply enclose it in a small polyethylene bag. Use a rubberband to hold it in place. The plastic wrap will have no effect on the detector's signal.

Take plenty of time choosing a search site. What you find depends to some extent on where you look. If you live in the New York metropolitan area and do your treasure hunting at Coney Island or Jones Beach, you'll find plenty of coins but they will almost all be contemporary ones. Switch operations to Central Park and the results will be different. Central Park has been heavily used by New Yorkers for more than a century, and coins found there sometimes have important value.

Find virgin territory, if you can. One summer morning recently, a pair of Massachusetts coinshooters, Al Dumais and Dick Boisseau, and their wives, drove their camper into Saint John, New Brunswick, a city of about 50,000, and pulled up by a park near the center of town. As soon as lunch was over, they started coinshooting. "We hit something with every sweep," says Dumais, "old coins, new coins, but hardly any junk. In about an hour our pockets were filled, and we had to go into the camper to empty them.

"Obviously, no one had ever come close to the park with a detector. We worked all afternoon, took a break for supper, and then went back again. We worked with flashlights once it got dark. It got to be almost midnight. Our arms ached; our backs ached. There was plenty more there, but we couldn't go on. I never saw anything like it."

A New Haven, Connecticut, coinshooter concentrates his energies on the eighteenth-century homes to be found in his area. Recently he came upon a 1792 home which was having an additional room built onto it. Checking the soil that had been excavated for the foundation, he found the following:

an 1842 large cent (Extra Fine)
an 1844 large cent (Fine)
an 1849 large cent (Extra Fine)
an 1857 large cent (Fine)
an 1858 silver three-cent piece (Fine)
an 1848 dime (Very Good)

Your own area is certain to offer a score or more of coinshooting sites. These are the places to try:

*Beaches* Beaches offer enormous potential, and the more popular the beach the more you're going to find in the way of lost coins and jewelry. (It's also true that you're going to find a greater number of beverage-can pulltabs, balls of crumpled up tinfoil, bottle caps, and other junk items.) In the case of public beaches, it's usually best to do your hunting in the early morning hours or on rainy days, that is, when the beach is not crowded. Otherwise, you may be overwhelmed by the curious.

When working in soft beach sand, many treasure hunters prefer to use a screen scoop instead of a garden trowel, screwdriver, or other digging instrument. Most detector dealers sell such scoops. When you use one, you're able to dig and sift in one operation.

When you arrive at the beach, the first thing to do is visualize how the beach is used, then plan your expedition accordingly. Some beachcombers concentrate on the area where people spread out their blankets and sunbathe. Notice where the refreshment stands are located. People are constantly handling money in this vicinity. At many beaches, benches are provided for nonbathers. Check beneath and around the benches. Scan under the boardwalks, too.

Often before the swimming season opens, the local parks department fills in areas that have been ravaged by winter storms with "new sand." It's not likely you're going to find any coins in such areas.

Other coinshooters prefer to work the area between the high tide line and the water's edge. There's much less junk there.

Check the condition of the tide before you set out. An ebb tide condition gives you the largest possible search area. Tide tables are usually carried in local newspapers or are available from marine supply shops and local Coast Guard facilities. You can also obtain tide tables by mail from the National Ocean Survey (Distribution Division, Riverdale, Maryland 20840). Specify whether you want the tables for the East Coast or those for the West Coast. Each set costs $2.

Keep in mind that tides are also subject to seasonal variations. So-called spring tides are those that cause the greatest rise and fall. They occur at or shortly after the new moon or full moon when the sun, moon, and earth are in approximate alignment. Experienced treasure seekers keep aware of when such tides are going to occur. They know that at such times the surf will be reaching high up onto the beach and in so doing will claim "new" coins and other objects.

A minority of beachcombers prefer to work right in the water, rather than on the beach proper. "Let the surf work for you," says one. "When the waves hit on the shore and the water then retreats, it acts like a net, dragging back small objects that have been buried in the sand. There are pockets where all this stuff collects. When you find one, you can recover several coins at one time."

If you do work in the water, you're never going to be impeded by frozen ground. You can treasure hunt all year round. Dress properly— long johns, wool pants, shirt and sweater; hip boots or waders—and you should be able to coinshoot on all but a handful of days.

If you're searching Florida beaches, or those in beach areas mentioned below, you'll want to get as close as possible to the water—or in it—with the idea of recovering gold or silver coins or other relics from a wrecked Spanish galleon. In the case of coins, don't look for something that is shiny and round. Salt water turns gold black, while silver and copper take on a dark greenish hue. Also as a result of the corrosive action of the salt water, coins lose their roundness. Look for dull, dark-colored, misshapen objects covered with residue and only vaguely of coin size.

The best time to do this type of hunting is immediately following a heavy storm. Wave action during the storm scoops up sand by the ton and churns it over and over. Thus, random coins are sometimes brought to the surface.

The beaches of Florida aren't the only ones on which ancient gold and silver coins are to be found. There's a twenty-mile strip on the western fringe of Delaware Bay extending north from Ocean City to Cape Henlopen, including Bethaney Beach, Dewey Beach, and Rehoboth Beach, where coins wash ashore. Off the coast lies the remains of the *Faithful Steward*, a vessel that foundered in attempting to enter the bay in 1785. In 1798, the *de Braak*, a British war sloop carrying treasure cargo pirated from Spanish ships, went down off Cape Henlopen. Professional salvage teams have had little success in recovering anything of value from these vessels because they lie in muddy waters, but they've provided many an exciting find for coinshooters.

The same is true farther to the north, along the beaches of south central New Jersey. On a private beach near Highlands, William Cottrell found a Portuguese "half Johanna," worth about $100. His son later picked up six more. When news of their good fortune leaked out, the beach was mobbed by treasure seekers raking, digging, and sifting. The site is probably still worthy of a search by someone equipped with a metal detector.

Rarely does a summer go by without someone reporting the finding of old items of trade goods being washed up on the Jersey shore. Beachcombers find spoons, rings, cuff links, buckles, pots, pans, and pewter and brass goods, all dating to the 1800s.

Old coins have been recovered on Massachusetts beaches, too, particularly on that strip of Cape Cod defined as the Cape Cod National Seashore Project, and, more particularly, between the towns of North Eastham and South Wellfleet. The coins recovered there are believed to come from the wreckage of the *Whidah*, a pirate vessel.

*Public Parks*  Search around drinking fountains, park benches, picnic tables, bandshells or near any other facility where people congregate. Picnic area and rest stops along the highways fall into this category.

*Playing Fields*  Beneath bleachers or grandstand areas is a logical place to look. Also around ticket booths and concession stands.

*Yards of Private Homes*  This suggestion refers to older houses, not development or tract structures, which are likely to have been built on newly filled land. When searching yards in established neighborhoods, give special attention to the grassy areas adjacent to sidewalks and any sites where children play.

*Playgrounds*  Search areas near playground equipment: swings, slides, climb-arounds, etc.

*Carnival Sites*  Amusement parks, circus grounds, or any other open area that offers—or once offered—recreational attractions of any type have potential. In most cases, it's a good idea to cover the site thoroughly from one end to the other.

*School Yards*  Ask yourself, "Where do the children play?" Then proceed accordingly.

*Parking Lots*  Naturally, this refers to *unpaved* parking lots. People getting in and out of automobiles and fumbling to find their keys or to pocket them, are bound to drop coins.

*Ski Areas*  Not in the winter, of course, but during the summer months when the snow has left and the crowds likewise, a ski slope can offer a variety of search sites. The best are likely to be where the tows load and unload, and at the end and the beginning of downhill runs. Coins are lost easily in the snow and virtually any spot where people gather offers the likelihood of several finds.

*College Campuses*  If you happen to live near a college campus, plan to spend some time searching there during the summer months, when most of the students aren't there. The more you know about the campus and the activities of the students, the more success you'll have. When do the students congregate? What sports do they play and where? Where are their favorite lounging spots?

*Railroad Depots*  Abandoned or not, commuter or otherwise, railway stations offer first-class possibilities. The area beneath and adjacent to waiting platforms should be scanned carefully.

Besides coins, you're sure to find medals, badges, tokens, and other such items. *(The Goldak Company)*

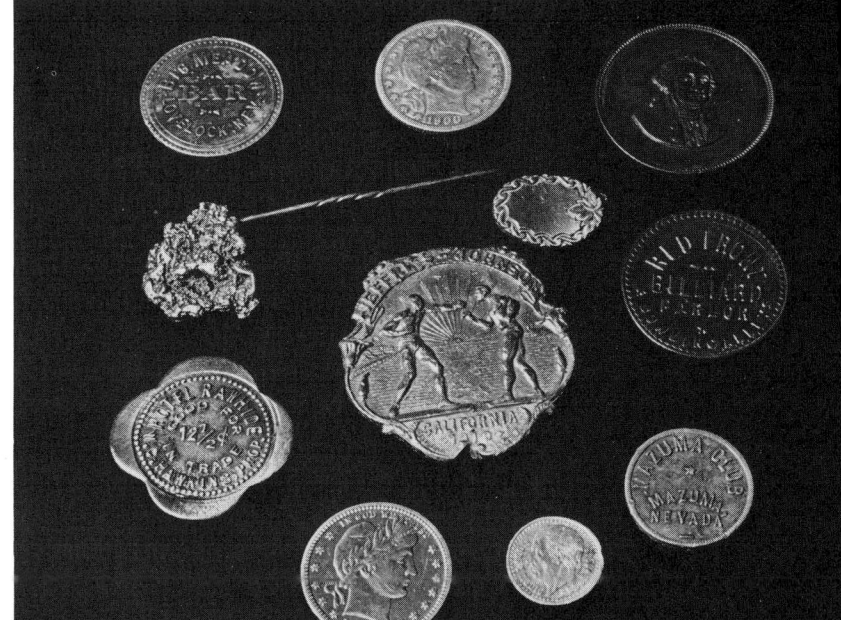

Gene Rolls of Forest Ranch, California, found these unusual coins and tokens (plus a gold-nugget stickpin) while coin-shooting. *(Ray Rolls)*

Don't feel that you have to keep finding new sites. A park that has been in use since the early part of the century will provide you with years of hunting fun. In your first searches, you're likely to recover the coins that are nearest to the surface. On subsequent trips, you may find fewer coins but they are likely to be better ones, older ones, those that are deeper down and give off fainter signals.

Use your imagination and you'll think of additional sites. In a recent issue of *The Treasure Hunter's Yearbook*, A. T. Evans told of a Kirkwood, Missouri, coinshooter who drives over the rural mail routes of his region; whenever he spots a mailbox he stops and searches beneath. He's found several hundred Indian Head cents. How come? Some time ago, he figured out that people put pennies along with unstamped envelopes into mailboxes, the coins are for the letter carrier to buy postage stamps. Sometimes, when the letter carrier reaches into the mailbox to remove the letters, the coins fall to the ground and are never recovered. This has been going on for the last century or so.

Evans gives a word of warning to go with his tip. "Don't open the mailbox," he says. "It's a federal offense."

Because much of your coinshooting is going to be done on playgrounds, on beaches, or in parks or involve other public places, you are usually going to have to contend with the public, with onlookers. Expect to get asked questions of all types. Here are some examples:

"What are you looking for?" (The most frequently asked question.)

"How does that thing the detector work?"

"Are you with the police?"

"Do you ever find money?"

"How do you know when you've found something?"

It usually pays to be straightforward, answering courteously and honestly. But be brief as possible, limiting your responses to one or two words whenever possible. Eventually an awkward silence will be created which will discourage the bystander. If you encourage conversation, you're going to put a damper on your success. Remember, you have to be able to concentrate.

Take solace in the fact that spectators today are a great deal less bothersome than they used to be. "You used to be looked upon as some kind of a freak," says Tom Powers, a veteran treasure seeker from Massapequa Park, New York. "But it's better now. Most people are aware of what you're doing, they leave you alone."

Sometimes onlookers can be helpful. They can tell you where people congregate in a park or describe the use patterns of a beach.

For a probe, use a screwdriver with a blunted tip. *(George Sullivan)*

In recovering a buried coin, first pinpoint exactly where it lies; probe into the ground.

Then cut a soil plug that includes the coin.

After recovery, replace the plug carefully and tamp it down. *(George Sullivan)*

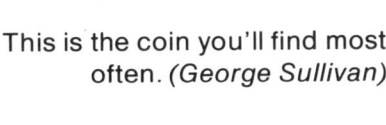
This is the coin you'll find most often. *(George Sullivan)*

## RECOVERING COINS

Once you've established where a coin is to be found, the idea then is to get it out of the ground quickly and with the least possible effort. If you're working on a park or on someone's lawn, you'll also want to dig in such a way that you don't do any permanent damage.

The first step is to pinpoint exactly where the coin lies. Go over the area several times with the search coil until you're sure of the exact spot to dig. Then probe the spot with a screwdriver, the tip blunted so it won't scratch the coin's surface. You're wasting your time if you do any digging without knowing the exact location of the coin. You can practice the art of pinpointing at home. Have someone conceal a dime or a penny under a magazine or a book. Then find where it is—exactly.

When you use the probe, don't jam it into the ground. Ease it in, then move the tip around until you feel it strike the coin. If the soil conditions are right, you can use the probe as a lever and propel the coin to the surface.

Charles Garrett, in his booklet, *Treasure Hunting Guide*, recommends a combination screwdriver-ice pick for successful coinshooting. The ice pick end of the instrument is to probe with; the screwdriver, to raise the coin. To make this tool, first remove the wooden handle from a standard ice pick. Take a screwdriver that's 8 to 10 inches in length and drill a hole in the handle end that is slightly bigger than the diameter of the ice pick. Fill the hole with epoxy cement and insert the ice pick. Using a grinder or file, blunt the ends of both the ice pick and the screwdriver. Last, put a slight bend in the screwdriver tip so that it will act somewhat as a scoop in raising the coin.

In hard-packed soil or thick turf, you'll have to use a tool that enables you to remove a cylinder-shaped plug of earth. A Bronx, New York, treasure hunter, who does most of his searching in the well-tramped soil of the parks of New York City, fashioned a digging instrument out of an old Army bayonet, cutting the blade down to 6 inches in length and then putting a new point on it. "It's rugged," he says, "and you can get a really firm grip on the handle." If you use such a tool, remember to pinpoint the location of the coin first using a blunted ice pick or screwdriver.

Many coinshooters follow the lead of Jim Alexander, a detector dealer in Houston and an avid coinshooter. Alexander tried using a shortened bayonet, but the blade was so thick that he couldn't push it into the rock-hard Texas soil during dry periods. A screwdriver would penetrate the soil all right, but he'd always tear up the soil getting the coin out.

Alexander's solution was to buy an inexpensive hunting knife with a slim 5-inch blade. He removed the plastic handle and, using an epoxy resin, replaced it with a 5-inch length of 1-inch wooden doweling. He then sanded down the dowel end.

Once he pinpoints the location of a coin, Alexander thrusts the knife blade into the soil at a slight angle, then carves a 3- or 4-inch circle around the coin's location. "It's like taking a plug out of a watermelon," he says. He then removes the plug from the ground using the knife blade. The plug has a cone shape.

To find out whether the coin is in the plug or still in the ground, he passes the loop over the plug. "Most likely," he says, "the coin is in your hand." Then he holds the plug over the hole and starts slicing half-inch sections from the bottom. Soon the coin falls free. He replaces what remains of the plug and taps it down with his toe.

Whatever type of tool you use, its efficiency should be such that it enables you to leave the search area exactly as you found it. You should be able to remove coins from a park lawn without anyone ever realizing the soil was disturbed. That's one hallmark of a master hunter.

## CLEANING AND STORING COINS

Treat any recovered coin with care. As soon as possible clean the coins using water and a household dishwashing detergent. This is vital in order to arrest metal corrosion that may have begun.

Use a tablespoon of detergent for each pint of water. A plastic bowl or pail should be used, not a metal container; the latter can react with the coin metal and cause additional damage. Soak the coins as long as necessary to loosen the surface dirt, then scrub each carefully with a soft toothbrush. Don't even use soap; it contains lye and ammonia, which can cause damage.

The final step is to rinse the coins several times in clean water, dry them, and then give them a final rinsing in isopropyl alcohol. Once dry, the coins are ready for checking and grading.

Be wary about doing any additional cleaning. If you want to enhance the appearance of a copper coin and heighten the inscription and date mark, you can rub it with a dab of cotton moistened in olive oil. Gold coins can be brightened with lemon juice. For silver coins, use the detergent solution described above. Keep in mind that anytime you rub or try to polish a valuable coin, you run the risk of marring or scratching the surface, thereby diminishing its value.

The surface of valuable coins should be protected by storing them in individual coin envelopes. Use a transparent envelope of some safe material, such as cellophane. Some coin dealers say that the popular 2 × 2-inch paper coin envelopes contain corrosive chemicals that can damage valuable coins. You can also use clear acetate holders to store and display your better coins, but these are expensive.

The cellophane envelope is only the first stage of protection. The coins should then be placed in ordinary paper coin envelopes. On each paper envelope, write the coin's denomination and the date it bears, and also information telling when and where it was found. Coin dealers sell especially designed boxes of tin or cardboard in which to store these envelopes. Each box holds as many as 200 coins.

## ESTABLISHING COIN VALUES

One fundamental reason why coinshooting is mushrooming in popularity is the ever upward drift of coin prices. Three reasons exist for this:

1. The inflationary trend of the economy. Like bread and Volkswagens, coins keep costing more.
2. The amount of coins most prized by collectors and dealers remains the same.
3. The number of collectors keeps increasing.

Add to this what has been happening in the precious metals markets in recent years. Both gold and silver sold at record prices in recent years; so did coins containing them.

There are occasional declines in coin prices, of course. Sometimes speculators, buying and selling certain issues in large quantities, send prices up, then quickly down. Or when huge reserves of coins suddenly come onto the market, as in the case of silver dollars recently released from Federal Reserve vaults, prices slump. But these actions affect the market only on a short-term basis. Within time, the upward trend resumes.

From a numismatic standpoint, two factors are of overriding importance in determining the value of almost any given coin. They are: availability and quality. The matter of the coin's age is important, too, but it doesn't have nearly the significance most people believe. I'm speaking now only of coins of the present century, the type you're going to be finding most of the time.

Availability relates directly to mintage, to the number of coins produced of a specific issue in a given year. Between the years 1939 and 1955, the lowest mintage nickel was produced in 1955, a fact that rendered that nickel more valuable than any of the others. The same holds true for the 1955 dime; fewer of it were minted than any other dime of the Roosevelt series and it's the most valuable.

One can easily find out how many coins of a given type were issued each year from the *Handbook of United States Coins,* an annual publication. Commonly called the Blue Book, it not only gives year-by-year mintage figures for each coin, it also lists the average amount dealers pay for each. The book costs $1.25. It's available from most coin dealers, or you can buy a copy by mail from the publisher, Western Publishing Co., Inc. (Whitman Hobby Division, Racine, WI 53404).

A companion book to this, titled, *A Guidebook to United States Coins,* usually referred to as the Red Book, also gives mintage figures, plus a listing of retail coin values. This book costs $2.50 and can be obtained in most coin shops or from the Western Publishing Co., Inc. (see address above).

Quality, a second important factor in establishing coin value, is much more a subjective matter than mintage. Collectors and dealers use eight different categories in describing coin quality, or lack of it. However, in assigning a grade to a specific coin, what is "very good" to you, the collector or seller, may be regarded as merely "good" or even "fair" by the dealer or buyer.

At any rate, these are the various categories:

*Poor (P)* A coin that is bent, corroded, or with its features or lettering and design almost completely worn away; undesirable for collectors.

*Fair (F)* A coin that is badly worn, with perhaps part of the lettering and design obliterated. Such coins are used by collectors as space fillers, that is, only until a better specimen can be obtained.

*Good (G)* This is the minimum standard acceptable to collectors and, hence, dealers. The date and mint mark are legible and the major features of the design discernible.

*Very Good (VG)* A serious amount of wear, but still valued by collectors. It is free of deep gouges or any other forms of mutilation.

*Fine (F)* Much of the lettering and detail show signs of wear, but still a desirable coin.

*Very Fine (VF)* Lettering and design are clear, but signs of wear are apparent.

*Extremely Fine (EF, XF, or E Fine)* All lettering and details of design are sharp and clear, although the very highest points of design show signs of wear.

*Uncirculated (Unc.)* A coin that has never been placed in general circulation. All lettering, the date, and design show in sharp detail. No scratches or mars. An absolutely perfect coin.

If you've had any contact with collectors or dealers, you've probably also heard coins described as being in "proof" condition. This term refers to coins that have been sold by a mint directly to a collector or dealer. They have a flawless surface and a mirrorlike sheen, and a close examination will show all design features to be extremely sharp in detail. It's not likely that you, as a coinshooter, will ever find a coin in "proof" condition.

## COINS OF THE GREATEST VALUE

Credit for finding what has to be one of the scarcest of all American coins has to go to Dan Cusanelli, a St. Louis restaurateur. Coinshooting near a big tree stump behind his restaurant, Cusanelli found a 1776 New Hampshire penny. It was buried six inches below ground level. Cusanelli's discovery rates not merely as an old coin. It's one of the oldest of all, *the* oldest, in fact.

Before the United States as such started minting coins, the individual states did. New Hampshire was the first to give consideration to the matter of coinage. On March 13, 1776, an act was passed in that state which provided for a committee to rule on the question of minting copper coins.

The committee empowered one William Moulton to make the coins. Moulton's "New Hampshire Coppers," as they are now called, are pure copper and size of present-day quarters and half dollars. Some show a pine tree; some a harp. A New Hampshire half penny has an evergreen of undetermined variety on the obverse and Moulton's initials and the inscription "American Liberty" on the reverse.

Vermont, Connecticut, and New Jersey also granted coinage privileges to companies or individuals. Massachusetts set up its own mint for copper coins. On rare occasions, treasure seekers in the Northeast do find examples of early state coinage but you can be sure good fortune plays a part in such strikes. These coins can bring anywhere from $10 to $200. It is unusual, however, for a person to sell a coin of this type. It becomes a family treasure.

It wasn't until March 3, 1791, that Congress passed a resolution calling for a mint to be established. Then followed with the passage of a second bill that provided that the money of the United States was to be based on the decimal system. The act specified coins of these denominations:

|                    | Face Value | Grains Standard |
|--------------------|-----------:|----------------:|
| Gold Eagle         | $10.00     | 270             |
| Gold Half Eagle    | 5.00       | 135             |
| Gold Quarter Eagle | 2.50       | 67 4/8          |
| Silver Dollar      | 1.00       | 416             |
| Silver Half Dollar | .50        | 208             |

| | | |
|---|---|---|
| Silver Quarter Dollar | .25 | 104 |
| Silver (Disme) | .10 | 41 3/5 |
| Silver Half Disme | .05 | 20 4/5 |
| Copper Cent | .01 | — |
| Copper Half Cent | .005 | — |

Through the years, the coining of some denominations was abandoned, new ones were added. Two-cent pieces, and three-cent pieces in two different varieties, were produced during the second half of the nineteenth century. And there were radical design changes in some coins. The cents that were minted from 1793 through 1858 were large ones, almost the size of today's half dollars.

The coins of greatest value for each denomination are cited in the pages that follow:

*Indian Head Cents* One of the most popular coins ever issued, the Indian Head cent is found with surprising regularity by astute coinshooters. The design for this coin was adopted in 1859 and mintage continued each year through 1909. Indian Head cents were minted of copper and nickel from 1859 through 1864; in the latter year the alloy was changed to 95 percent copper and 5 percent zinc. Coins of this type were produced until 1909, the year the Indian Head was discontinued. The first branch mint cents were produced in San Francisco in 1908.

Virtually any Indian Head cent produced before 1880 has important value, although the matter of condition is crucial. For instance, an 1866 cent in Good condition has a catalog value of $7.50, while the same cent is listed at $55 when condition is Extremely Fine.

Other Indian Head cents of important value include: 1885, 1886, 1894, 1908-D and 1909-S. The Indian Heads for 1889 through 1891, and 1893, have the least value, but even these are worth 40¢ when in Good condition and $5 to $6 when Extremely Fine.

*Lincoln Cents* This coin was first issued in 1909 to commemorate the 100th anniversary of Lincoln's birth. It bore a portrait of Lincoln on the obverse, and ears of wheat on the reverse, thus collectors sometimes refer to these coins as "wheaties." In 1959, in honor of the 150th anniversary of Lincoln's birth, the design of the reverse was changed to a representation of the Lincoln Memorial.

Many cents are unusual for one reason or another, and these are the ones to look for. Both the 1909 and 1909-S bear the initials V.D.B. on the bottom of the reverse, which stand for Victor David Brenner, the designer of the coin. The 1909-S, in merely Good condition, is worth almost $100.

The most valuable Lincoln cent of all was produced in 1909. The Denver mint turned out 7,160,000 cents that year, while the Philadelphia mint didn't produce any. But none of the cents were made to bear the Denver mint mark, a D. The coin has sold for as much as $1,000. Other Lincoln cents of particular value are 1911-S, 1914-D, 1914-S, 1915-S, 1924-D, 1926-S, 1931-D, 1931-S, and 1933-D. As in the case of Indian Head cents, condition is a critical factor.

Most treasure seekers covet the 1943 cent because of its silver-gray coloring. As the result of a critical copper shortage during this war year, the Treasury Department used zinc-coated steel for all cents. But 1943 cents have yet to obtain high value because of the huge quantities minted. The 1943 and 1943-D cents, when in Fine condition, are listed in catalogs at 15¢, while the 1943-S, also in Fine condition, is worth 30¢.

Any cents issued from 1944 to the present must be at least Very Fine or Extremely Fine in condition to have any value more than face value, with the exception of the 1955 cent. The dies from which some coins of this date were minted were improperly prepared, the result being that the date and the inscriptions "Liberty" and "In God We Trust" appear in double outline. These double-struck cents are worth nearly $200 in Fine condition.

Cents are produced in enormous quantities nowadays. More than 3 billion of the 1964-D were minted, and more than 4 billion of 1969-D.

Despite the preceding remarks, the Lincoln cent had a lowly status until 1974. That was the year that George P. Schultz, then Secretary of Treasury, asked Congress to grant his department authority to produce the aluminum pennies. The problem was that the price of copper was approaching the point where the metal content of the coin was worth more than one cent.

Mint officials had been watching the escalating price of copper with increasing dismay, knowing that if the rise continued, mass hoarding of pennies would be the result. In March 1974, the price of copper hovered around $1 a pound. If the price were to go to $1.50 a pound, hoarding would become widespread, said Mint officials. At $1.20 a pound, it costs the Mint more than a cent to make a penny.

Copper is traded on the New York Commodity Exchange. You can consult the *Wall Street Journal* or the financial pages of the *New York Times* or most any big city daily to find out the current price.

The aluminum penny would contain 96 percent aluminum and 4 percent of some other metal, magnesium is the most likely candidate. The Mint has indicated there is to be no recall of copper pennies. About 34 billion were in circulation in 1974, with the average life ranging from 20 to 25 years.

The Indian Head cent.
*(American Numismatic Society)*

The Liberty Head nickel.
*(American Numismatic Society)*

The Mercury dime.
*(American Numismatic Society)*

Standing Liberty quarter.
*(American Numismatic Society)*

Liberty Walking half dollar.
*(American Numismatic Society)*

*Nickel Five Cent Pieces* Jefferson Nickels, first produced in 1938, are the ones you're going to find with the greatest frequency. Except for 1939-D and 1950-D, these coins were produced in large mintages. Some worthwhile items in the series include: 1938-D, 1939-S, 1949-S, 1950-P, and 1951-S.

Coins produced in 1942, 1943, 1944, and 1945 have particular value too. These were wartime years and nickel was short in supply, so the Jefferson coin was minted in a special alloy that consisted of 56 percent copper, 35 percent silver, and 9 percent manganese. A nickel that is 35 percent silver is worth, by 1974 standards, at least six or seven times its face value.

To distinguish the 1942-1945 coins from those that preceded them, the mint mark was enlarged and placed over the dome of Monticello on the reverse. A "P" was used as the Philadelphia mint mark, which had never been done before.

As just about everyone over forty knows, the Jefferson nickel replaced the Indian Head or Buffalo Type nickel, which was produced from 1913 to 1938. Indian Heads were minted of a copper-nickel alloy and were actually 75 percent copper.

Several of the earlier dates of this issue have attained high values. For example, the 1914-D, in good condition, is listed at $16 in some catalogs. The 1915-S, 1921-S, and 1926-S are other Indian Heads that are highly prized by collectors.

The Liberty Head nickel was produced from 1883 through 1913. These were of the same composition as the Indian Head nickel, 75 percent copper, 25 percent nickel. Any Liberty Head nickel in Good condition, and produced between 1900 and 1912 when mintages were generally the highest, is worth from 50¢ to 75¢. Virtually all other Liberty Heads are listed at several dollars each.

Before the Liberty Head nickel went into circulation, the Shield Type five-cent piece was in use. Only rarely do coinshooters report finding these.

*Dimes*   Beginning in 1873, the weight of the dime was established at 2.50 grams and its composition at .900 silver. Dimes were minted to these specifications until 1965, the year the weight of the coin was reduced to 2.27 grains, while its value was whittled back to, well, 10 cents, since this was the year that silver coinage gave what the Treasury Department calls "clad coinage," evidenced by those sandwiches of copper and nickel that now make up your pocket change.

The Roosevelt dime, the type in circulation today, was first minted in 1946. There are no real rarities in this series yet, but in some years the mintages have been relatively low. For example, less than 21 million of 1950-S were produced, as compared to 1960-D and its slightly more than 200 million. And in the clad coinage year of 1967, more than 2 billion dimes were minted.

Coinshooters frequently find the Winged Liberty Head dimes, those commonly referred to as Mercury dimes. These preceded the Roosevelt dime in circulation, being minted from 1916 through 1945. In the case of the 1916-D Mercury, only 264,000 were minted. When in Good condition, it lists at $95; at $200 when Fine. Other Mercury dimes that have notable value are: 1921, 1921-D, 1926-S, 1930-S, 1931-D, and 1931-S.

The Liberty Head dime was produced from 1892 through 1916. Designed by Charles E. Barber, chief engraver of the mint, the coin is often referred to as the "Barber dime." All dimes in this series have high value today. The minimum catalog listing for a Liberty Head dime is 65¢, when condition is Good, and for a number of years the value is pegged at over $20: 1894-O, 1895, 1895-O (with a mintage of only 440,000), 1896-O, 1896-S, 1897-O, and 1901-S.

*Quarter Dollars*  Up until 1965, when clad coinage came into being, quarters weighed 6.25 grams and were 90 percent silver. Based on the price of silver in February, 1974, this means that any quarter bearing a 1964 date or earlier is worth almost three times its face value.

The Washington quarter, the one in circulation today, was first issued in 1932. Be on the lookout for 1932-D and 1932-S, the most valuable coins of the series. In Good condition, they are worth between $30 and $40. Other Washington quarters of value include 1934-D, 1936-D, 1937-S, 1938-S, 1939-S, 1940-D, and 1955-D.

The Standing Liberty quarter, minted between 1916 and 1930, disappeared from general circulation many years ago. Anytime you recover one that is dated 1924 or before, you've made a real find. The 1919-D, 1919-S, 1920-D, and 1923-S are especially scarce. In Good condition, any of these brings $10 to $20.

Before the Standing Liberty quarter, we had the Liberty Head, which was minted from 1892 to 1916. Like the dime, it was designed by Charles E. Barber, and is often called the "Barber quarter." The specifications for the Liberty Head—6.25 grams; composition, .900 silver—were followed in the Standing Liberty quarter (up until 1965). All Liberty Head quarters have good value and for some the demand is exceptional, including 1896-S, 1901-S (of which only 72,664 were minted), and 1913-S.

From 1839-1891, the Treasury Department circulated Liberty Seated quarters. These weighed 6.68 grams, slighty heavier than the Barber quarter and the others that were to follow, but with the same fineness. Quarters of this type, in Good condition, bring $4 or $5 at the very minimum.

*Half Dollars*  Coinshooters realize that they have about as much chance of recovering a Kennedy half dollar, the half dollar in circulation today, as they do of spotting a half dime or a silver three-cent piece. Even though the Treasury Department turns out Kennedy halves in huge quantities—450 million were produced in 1971 alone—people keep stowing them away. When, in 1964, the Kennedy half dollar was first issued, it weighed 12.50 grams and was .900 silver. It stayed that way for only one year. From 1965 through 1970, the coin was merely clad in silver, and in 1971 its composition was changed to a copper-nickel alloy. It doesn't matter what the Kennedy half happens to be made of, however, people hoard every type.

From 1948 through 1963, the Franklin half dollar was circulated. A total of 35 different varieties were produced. In the case of 10 of the 35 pieces, the mintage figures were less than 5 million, so, as you may judge, values are already quite high. Unfortunately, many Franklin halves were only lightly struck, and thus lack in detail, which diminishes their value. Scarce dates for the Franklin half dollar are 1949-S, 1953, and 1955.

It is one of the anomalies of coinshooting that the half dollar most frequently found is the Liberty Walking half, which hasn't been minted since 1947. Weighing 12.50 grams and composed of .900 silver, the Liberty Walking half was first produced in 1916. The reason for its prevalence is that, unlike the halves of today, it was a standard piece of pocket change, desired only for its face value. Several coins of this series have excellent values, including 1916-S, 1921, 1921-D (of which only 208,000 were minted), 1921-S, and 1938-S.

A Liberty Head half dollar was minted from 1892 through 1915. Like the dime and quarter, the coin was designed by Charles E. Barber. Its specifications were the same as the Liberty Walking half. Any Liberty Head half dollar, in Good condition, catalogs for a minimum of $3. Several coins in the series exceed $30 in value: 1892-S, 1893-S, 1896-S, and 1897-S.

*Silver Dollar* Among the first coins to be authorized by Congress in April, 1792, but not produced until 1794, the silver dollar is probably sought more avidly than any other coin. The reason isn't hard to understand. The earliest silver dollars weighed 416 grains and were .892 fine. The weight was reduced in 1837 to 412½ grains, while the fineness was increased to .900. Thus, there can be no quibbling about the value of this coin. It has real heft to it. And it's *silver*.

Random finds of silver dollars are rare, unfortunately. When coins of this denomination do happen to be recovered, it's usually as part of a treasure cache of some kind. Recently an Iowa man found 117 silver dollars in a rusted tin box beneath the front steps of an old farmhouse that had once been occupied by his grandparents. It was not exactly a lucky strike; he had an idea that grandpa had hidden coins someplace.

Silver dollar coinage lapsed from 1804 to 1840, and again from 1873 to 1878. Silver dollars were discontinued after 1935. When production resumed in 1971, the only thing silver about the coin was its appearance, since it was minted of a copper-nickel alloy.

The design of the silver dollar has changed many times. The best known is the Peace Dollar, which first appeared in 1921 and was produced each year through 1935. The most valuable Peace Dollars are the 1928, worth $75 when Very Fine, 1921, 1927, 1927-D, 1927-S, 1934, and 1934-S.

The silver dollar first issued in 1971 honored President Dwight Eisenhower. While circulation issues of this coin are copper-nickel in content, the Treasury Department has also produced special silver issues for collectors.

The Liberty Head silver dollar was produced from 1878 through 1921. One George T. Morgan designed the coin, thus, collectors sometimes refer to it as the Morgan Dollar. Some of these coins have exceptional value. The 1893-S, in Very Fine condition, lists at $300, and the 1889-CC at $115, also when Very Fine.

From 1840 through 1873, the silver dollar featured a seated figure of Liberty on the obverse. Any coin of this series, in merely Very Good condition, is worth in the neighborhood of $50, and some Liberty Seated types bring ten times that amount.

*Commemorative Coins* If you're a dedicated coinshooter, skilled in how to search and knowledgeable about location, you're occasionally going to find a commemorative coin. While it's true that most commemoratives are to be found in coin dealers' showcases or collectors' safe-deposit boxes, a number of issues managed to find their way into general circulation and once were considered pocket change by most people.

These special coins, which usually took the form of half dollars, were issued to commemorate special events, help pay for historical monuments, or memorialize people important in American history. Each suggestion for a commemorative coin was considered individually by the appropriate Senate and House committees, and then minted once Congress had given its approval.

The first United States commemorative coin was issued in 1892 in connection with the 400th anniversary of Columbus's voyage to America and the Columbian Exposition in Chicago which took place the same year. You can buy these coins today for about $10. In the half century following the minting of this first commemorative, approximately 50 different other types were issued. Counting the mint mark varieties, there have been a total of 142 coins in the series. None were struck from 1940 to 1945, and none since 1954.

Until recent years and the escalating interest in coins and precious metals, the general public had little knowledge of commemorative coins. Once minted, they would be sold to dealers and then to collectors, always in bright, uncirculated condition. Some were extremely popular. But some weren't, and the quantities minted were never sold. The unsold coins were usually returned to the mint to be melted, but in some cases the returned coins were placed into circulation.

If you do happen to find a commemorative, it is likely to be one of these. A large but unknown quantity of Columbian half dollars was released into circulation, as were quantities of the Stone Mountain Memorial (1925), the Monroe Doctrine Centennial (1923), and the Pilgrim Tercentenary (1920).

Circulated commemoratives of more recent mintage include the Booker T. Washington Memorial issued over a period of six years beginning in 1946, and the Washington-Carver half dollar which honored prominent Negro Americans, Booker T. Washington and George Washington Carver. Minted in 1951 through 1954, this coin was meant to provide funds "to oppose the spread of communism among Negroes in the interest of national defense."

Commemorative half dollars have a wide range of values. Some, such as the Washington-Carver and Booker T. Washington commemoratives, are valued more for their silver content than anything else. At 1974 prices, they were worth $1.35 apiece. But other commemoratives are much prized by collectors, and there are several different issues worth well over $100. Not all commemoratives were silver. Thirteen issues were minted of gold and these, of course, have enormous value.

In the case of silver commemoratives, to establish the value of individual issues, first find out how many coins of the issue were minted, then determine how many, if any, were returned to the mint and melted, and, finally, how many happen to be in the hands of dealers and collectors. Any good coin book (see below) contains this information.

## COINS AND SILVER

The activities of the precious metals market have complicated the task of establishing coin values. During the early 1960s, the price of silver, which the Treasury Department had held stable at 92¢ an ounce for years, was permitted to rise to $1.30 an ounce. The idea was to protect the Treasury reserves from depletion. The government also halted the production of Silver Certificates and, instead, issued Federal Reserve notes.

But silver reserves continued to dwindle, and by 1964 it was apparent that the Treasury Department would have to halt production of coins that were mostly silver. With the rise in the price of silver, millions of Americans (whose only previous interest in coins was knowing whether they had the right change to make a telephone call or pay a highway toll) suddenly became "collectors," lapping up virtually all the silver coins in circulation.

Faced with a serious coin shortage, the Treasury Department ended the minting of silver coins with the Coinage Act of 1965. Under the provisions of the Act, half dollars, formerly 90 percent silver, were to have only 40 percent silver content. Quarters and dimes were to be minted of copper and nickel, with no silver at all. This resulted in the coin "sandwiches" common nowadays, the dimes and quarters with a copper-colored stripe around the outside edge.

Go through your pocket change for a week, and if you find a silver coin, consider it remarkable. They're all being hoarded. Coinshooters, however, find about as many pre-1964 coins, mostly silver, as they do recently minted ones.

In establishing the value of silver coins, the factors of availability and quality are not as critical as they once were. The factor of silver content overshadows everything else. There are some exceptions, of course, but they're becoming fewer and fewer.

During 1974, the price of silver on the New York Commodity Exchange rose to within a few cents of $4.70 an ounce. This meant that a silver dime—no matter the number minted, no matter the condition—was worth about 33¢. A quarter was 82¢; a half dollar, $1.64, and a silver dollar, $3.52.

This chart shows how the price of silver affects the value of coins that are mostly silver:

| Price of Silver (per ounce) | Value of Silver Content | | | |
|---|---|---|---|---|
| | Dime | Quarter | Half Dollar | Dollar |
| $1.30 | $ .09 | $ .24 | $ .47 | $1.00 |
| 1.40 | .10 | .25 | .51 | 1.01 |
| 1.50 | .11 | .27 | .54 | 1.16 |
| 1.60 | .12 | .29 | .58 | 1.24 |
| 1.70 | .12 | .31 | .61 | 1.31 |
| 1.80 | .13 | .33 | .65 | 1.39 |
| 1.90 | .14 | .34 | .69 | 1.47 |
| 2.00 | .14 | .36 | .72 | 1.55 |
| 2.10 | .15 | .38 | .76 | 1.63 |
| 2.20 | .16 | .40 | .80 | 1.70 |
| 2.30 | .16 | .40 | .83 | 1.84 |
| 2.40 | .17 | .43 | .87 | 1.86 |

| Price of Silver (per ounce) | Value of Silver Content | | | |
|---|---|---|---|---|
| 2.50 | .18 | .45 | .90 | 1.94 |
| 2.60 | .19 | .47 | .94 | 2.01 |
| 2.70 | .19 | .49 | .97 | 2.09 |
| 2.80 | .20 | .51 | 1.01 | 2.17 |
| 2.90 | .21 | .53 | 1.06 | 2.26 |
| 3.00 | .22 | .54 | 1.09 | 2.32 |
| 3.10 | .22 | .56 | 1.12 | 2.40 |
| 3.20 | .23 | .58 | 1.16 | 2.47 |
| 3.30 | .24 | .59 | 1.19 | 2.56 |
| 3.40 | .25 | .61 | 1.23 | 2.63 |
| 3.50 | .25 | .63 | 1.26 | 2.72 |
| 3.60 | .26 | .65 | 1.30 | 2.78 |
| 3.70 | .27 | .67 | 1.34 | 2.86 |
| 3.80 | .27 | .69 | 1.37 | 2.94 |
| 3.90 | .28 | .70 | 1.40 | 2.99 |
| 4.00 | .29 | .72 | 1.45 | 3.09 |
| 4.10 | .29 | .74 | 1.48 | 3.18 |
| 4.20 | .29 | .76 | 1.51 | 3.26 |
| 4.30 | .30 | .77 | 1.55 | 3.33 |
| 4.40 | .31 | .79 | 1.58 | 3.41 |
| 4.50 | .32 | .81 | 1.62 | 3.49 |
| 5.00 | .35 | .90 | 1.80 | 3.88 |
| 5.10 | .36 | .92 | 1.84 | 3.95 |
| 5.20 | .36 | .94 | 1.87 | 4.03 |
| 5.30 | .37 | .95 | 1.91 | 4.11 |
| 5.40 | .38 | .97 | 1.94 | 4.19 |
| 5.50 | .39 | .99 | 1.98 | 4.26 |
| 5.60 | .39 | 1.01 | 2.02 | 4.34 |
| 5.70 | .40 | 1.03 | 2.05 | 4.41 |
| 5.80 | .41 | 1.04 | 2.08 | 4.49 |
| 5.90 | .41 | 1.06 | 2.12 | 4.56 |
| 6.00 | .42 | 1.08 | 2.16 | 4.65 |
| 6.10 | .43 | 1.10 | 2.20 | 4.73 |
| 6.20 | .43 | 1.12 | 2.23 | 4.81 |
| 6.30 | .44 | 1.13 | 2.27 | 4.88 |
| 6.40 | .45 | 1.15 | 2.30 | 4.96 |
| 6.50 | .46 | 1.17 | 2.34 | 5.04 |
| 6.60 | .46 | 1.19 | 2.38 | 5.12 |
| 6.70 | .47 | 1.21 | 2.41 | 5.19 |
| 6.80 | .48 | 1.22 | 2.45 | 5.27 |
| 6.90 | .48 | 1.24 | 2.48 | 5.35 |
| 7.00 | .49 | 1.26 | 2.52 | 5.43 |

## COINS AND GOLD

What about gold coins? An Indianapolis police officer recently found an 1887 five-dollar gold piece in a local park, but this is very much an exception. Coinshooters find gold coins only on the rarest of occasions.

Nevertheless, what's happening in the gold-coin market is worth noting. During the early months of 1974, coin dealers were wary about quoting prices on gold coins because of the escalating cost of gold. One major dealer advised prospective customers that the prices being quoted in his company's advertising were only "reference prices."

On some days during February 1974, the values of popular American gold coins—the double eagle ($20 piece), eagle, and half eagle—changed from hour to hour. A double eagle that sold for $70 in 1970 cost $300 that month. Despite the mushrooming prices, customers crowded coin stores in New York. The *New York Times* described dealers as being "swamped" with buyers.

## METAL TOKENS

If you have the diligence and patience to find coins, it's inevitable you are also going to be finding metal transportation tokens. Collecting them has become a hobby in its own right. A rare token sells for as much as $10 or $15, even more.

Transportation tokens are stamped out of brass, lead, nickel, or tin. Most are round and about the same size as a dime, although there are many quarter-sized. Some are oval, others square, and still others octagonal. Most tokens that you will find were purchased by transit patrons for 10 to 25 cents.

Treasure hunters have recovered tokens dating to the 1840s. Experts say that more than 8,000 different tokens have been issued.

To students of the history of transportation in the United States many of the tokens have important value, for in many cases they represent transportation enterprises that have gone the way of the buggy whip. In the big cities of the Northeast, tokens were issued for horsecar lines. Cities everywhere issued trolley-car tokens. Boat lines up the Hudson River had their own tokens, and scores of railroad lines, who were to merge with the New York Central, the Pennsylvania Railroad, or other giants, are remembered through tokens and little else.

Some coin collectors scoff at those who specialize in tokens. It is proper, however, that these items be classified as numismatic material, for there are any number of examples of tokens being used as money. In New York City, merchants and newspaper dealers will give customers a subway token as a portion of their change. It's a convenience; the customer doesn't have to wait in a token line. In Buffalo, New York, transit tokens were once in general use as small change.

Tokens sometimes have many of the same characteristics as coins. Occasionally they will be dated, and some have distinctive designs. The Oakland, Brooklyn & Fruitvale R.R. Co., a California transit system, issued a token which pictured a horse pulling a horsecar, while the Des Moines (Iowa) Railway Company's token bore a design showing a trackless trolley; that is, a bus that derived its power from overhead electric wires. Many tokens had a simple letter design: W for Washington, D.C., and S for San Diego are two examples.

Another type of metal token you may see occasionally is the sales and tax token. Arizona, Colorado, Illinois, Kansas, Louisiana, Missouri, New Mexico, Oklahoma, Utah, and Washington are among the states that have issued them. Compared to transit tokens, they are newcomers to the field, having first been issued in the 1930s.

Many of these tokens were issued in denominations of from one to five mills, a mill being a monetary unit equal to 1/1000 of a dollar. Often they were made of either copper or aluminum. There was usually not the slightest effort to achieve artistic merit in the design of these tokens. Some bore only a legend, such as "Missouri Sales Tax Receipt," "Luxury Tax Token" (Alabama), or "Emergency School Tax on Purchases of 5 cents" (Utah).

Although such tokens were issued by the millions, they were of such miniscule value that people treated them with little regard. As a result comparatively few of them are in existence today. As a general rule, however, they are less valuable than transit tokens.

If you become interested in tokens, it will not be long before you begin to hear and read about a wide array of nineteenth-century privately issued tokens which were known variously as copperheads, Civil War tokens, Hard Times tokens, Jackson tokens, or store-cards (because they bore a store's name and address). These were cent-size and quarter-size, usually made of copper, but also of silver, lead, and zinc. They came into use during periods of American history when people took to hoarding government-minted coins. More than ten thousand different varieties of these tokens have been identified and random finds are being made all the time. A Bronx, New York, photographer recently recovered an 1837 Hard Times token in Van Cortlandt Park, not far from his home.

The Director of the Mint came to regard all of these tokens as unauthorized substitutes for coins, and laws were passed in 1864 to prohibit private coinage of any kind. Despite their historical appeal, and the fact that these tokens are all at least a century old, they have never attained great value. The best of them trade back and forth for just a few dollars.

Metal tokens of many types are issued today, most of them as sales promotion devices. In the late 1960s, the Shell Oil Company distributed a series of 35 different quarter-size tokens, each bearing the likeness of a president, to boost patronage at the company's service stations. Sunoco sponsored a similar promotion at about the same time, distributing aluminum tokens bearing pictures of antique automobiles. To future generations of a fuel-short world, such tokens are sure to appear anomalous and, therefore, have value as curios.

Coin publications and books about coins often contain information on historical tokens of various types. But the best way to get more information on the subject is by joining a token collectors club or organization.

The American Vectorist Association (P.O. Box 31, Clinton, CT 06413).
Civl War Token Society (8024-N South Mulligan Ave., Oak Lawn, IL 60459).
Token and Medal Society, Inc. (P.O. Box 82, Lincoln, MA 01773).

## FOR MORE INFORMATION

As your collection builds, you'll want to increase your coin knowledge and perhaps investigate different methods of selling what you've found in either the wholesale or retail market. A number of periodicals are devoted to coins and collecting. Write and request sample copies. The publications include:

*Coins Magazine* (Iola, WI 54945). One year, 12 issues, $4; sample copy, 50¢.
*Coin Dealer's Newsletter* (Box 2308, Hollywood, CA 90028). One year, 52 issues, $10; sample copy, 25¢.
*COINage Magazine* (16250 Ventura Blvd., Encino, CA 91316). One year, 12 issues, $4.75; sample copy, 60¢.
*Coin World* (Box 150, Sidney, OH 45365). One year, 12 issues, $3; sample copy, 25¢.
*Numismatic News* (Iola, WI 54945). One year, 26 issues, $6; sample copy, 25¢.
*The Numismatist* (P.O. Box 2366, Colorado Springs, CO 80901). One year, 12 issues, $7.50; sample copy, 75¢.
*Numismatic Scrapbook* (P.O. Box 150, Sidney, OH 43565). One year, 12 issues, $5; sample copy, 50¢.

Two excellent books, informative and authoritative, are:

*The Complete Book of Coin Collecting.* Joseph Coffin. New York: Doubleday & Co., 1970; $5.95.

*The Complete Book of United States Coin Collecting.* Norman Davis. New York: The Macmillan Co., 1971; $6.95.

You can also build your knowledge of the field by joining a collector's organization. The American Numismatic Association (818 North Cascade Blvd., Colorado Springs, CO 80901) is recommended. This organization conducts research on coins, coinage, and the history of money, and publishes articles on these topics in its monthly magazine, *The Numismatist*. The ANA also maintains a library and museum and holds annual conventions.

About one hundred local coin clubs are affiliated with the ANA. A list of them is published periodically in *The Numismatist*.

Also investigate the American Numismatic Society (on Broadway between 155th and 156th Sts., New York, NY 10032), an organization dedicated to "the advancement of numismatic knowledge." It maintains one of the most comprehensive numismatic libraries in the world and a fascinating museum.

## Chapter 5

# WHAT ELSE TO LOOK FOR

|                       |                      |
|-----------------------|----------------------|
| Cowbells              | Cast-iron toys       |
| Sheep bells           | Spoons, knives, forks|
| Spurs of solid brass  | Horseshoes           |
| Kerosene lanterns     | Stove doors          |
| Cigarette lighters    | Drum majors' batons  |
| Cartridge cases       | Keys                 |
| Horses' bits          | Nails                |
| Sailmakers' thimbles  | Eyeglasses           |
| Pewter sugar bowls    | Ice tongs            |
| Meat cleavers         | Ice skates           |
| Buttons               | Aluminum saltcellars |
| Survey stakes         | Monkey wrenches      |
| Railroad spikes       |                      |

If you were to use a detector every day for a week, searching local parks, playgrounds, and beaches, and you made a list of the items that you recovered, their number and variety would surprise you. Some of the objects you would find are included in the list above.

Most of what you find will be worthless, but you're bound to find some valuable items, particularly if you're careful in selecting a search site.

If you haven't visited a flea market or antique shop recently, you may not be aware of the many different things that have value today. These aren't necessarily antiques, an antique being, according to the U.S. Tariff Act of 1930, "a work of art, works in bronze, marble, terra cotta, porcelain, pottery, parian, and artistic antiquities and objects of ornamental character or educational value...*produced prior to 1830.*" You may find an antique or two, but you're more likely to find collectibles. These are objects which, because of their beauty, rarity, oddness, or significance in popular culture have intrinsic value to a collector. Their date of manufacture is not of particular significance. Jim Beam bottles, a type of whiskey bottle, are examples of collectibles. They are eagerly sought by collectors and dealers. Jim Beam bottles were produced during the early 1950s.

The pages that follow examine some of the categories of collectibles in which treasure hunters specialize.

## BOTTLES

Treasure hunter John Turner, searching the ghost town of Prospect, Nevada, discovered what was once the dumping area for the Happy Times saloon. Beneath a heavy covering of weeds and a thin layer of soil, he recovered a hoard of bottles which have been appraised at more than $7,000.

Bottles are collected by thousands of treasure hunters, sometimes to the exclusion of everything else. Usually they're found in old dumps. The detector is able to seek them out because of their proximity to old tin cans or other buried metal items. Sometimes a jar's zinc top is what sets off the detector.

Bill Schreiber of Gaylord, Michigan, is a bottle specialist. In his first year as a treasure hunter, he was highly successful in finding old bottles on the farms and backwoods logging camps on the northern stretches of Michigan's Lower Peninsula. He sold most of the bottles he found and used the money to buy a more sophisticated detector.

In his second year of hunting, he found many more bottles, and often they were of higher quality. He repeated his strategy, selling off most of what he found and purchasing a deluxe detector this time, one with a gold probe and a headset. Now in his third year of bottle collecting, he has more than 500 different specimens and he's planning his next move.

What makes a bottle valuable? Collectors look for specimens that are as close to mint condition as possible. Scratches, cracks, or stains reduce the value, but bubbles are not considered detrimental. Embossed designs or unusual names or phrases increase the bottle's worth.

Early stoneware bottles were among those found by this collector.
*(George Sullivan)*

Age has a great deal to do with it. Some bottles are very old.

The history of glassmaking in America has been traced to the earliest colonists. When the settlement of Jamestown was founded in 1608, it included a glassmaker's house. Located about a half a mile from the main clusters of homes, it has come to represent the first industry established in the New World. There is evidence that the small Jamestown factory turned out table glass and bottles which were shipped to England and sold there. Ultimately, the enterprise failed.

What is often called the first successful glasshouse in America was founded in 1739 by Caspar Wistar, a Philadelphia button-maker. His factory in Salem County in southern New Jersey, employing glassmakers imported from Germany, turned out free-blown bottles of assorted sizes, many of them graceful in shape and attractively ornamented. Before the end of the eighteenth century, other glasshouses were in operation. One in Massachusetts, for example, supplied cider bottles that were used to ship the product to the West Indies.

Henry William Stiegel was another famous glassmaker of this period. He brought over glassmakers from Germany and England and established a town for them in Pennsylvania which he called Mannheim. His factory produced a variety of glass objects which were the equal of the best being manufactured in Europe. But despite the high quality of what he produced, Stiegel's business failed in 1774.

After the Wistar and Stiegel factories had closed, there was practically no glassmaking in the colonies, except for the production of windowpanes. Bottles, decanters, and other glass products had to be imported from England. That's the way the British wanted it. They meted out severe punishment to any glass blower who sought to leave England and establish his trade in the colonies. After the Revolutionary War, with England no longer supplying America's needs, glassworks began to spring up everywhere. Often they were staffed by artisans who had worked for Stiegel.

Close to forty glass factories were in operation by 1800, all making blown glass of varying quality. Factories were located in Philadelphia, Boston, and Pittsburgh, and in the smaller communities of Massachusetts, Connecticut, New Hampshire, Vermont, New York, and Ohio.

Bottles of the period were of many shapes and sizes. Often they were used for whiskey, gin, and other spirits. They were usually rough and irregular in form. The mouth did not have a smooth rim, merely a rough-edged opening.

Bottles of the early nineteenth century are often identified by their pontil marks. The pontil was a long iron rod used to take the blown bottle from the blowpipe, an operation that left a rough scar on the finished product.

The horse was the chief form of transportation in those days, and the rider, to carry whatever he drank, needed not a bottle in the usual sense but a "pocket bottle," a flask. Small bottles that were flat-sided and short-necked were in constant demand throughout the early decades of the nineteenth century. You will not be a bottle collector for very long before you hear of the Pitkin flask, the generic name for the olive-amber, pint-size flask, the first of which were maufactured at the Pitkin Glassworks in East Hartford, Connecticut, and later in Manchester.

Later there were portrait flasks. George Washington's Portrait appears on at least sixty different flasks, and Benjamin Franklin, Thomas Jefferson, John Adams, Andrew Jackson, and Henry Clay were similarly honored. Other flasks were decorated with the American eagle, the flag, Columbia, Masonic emblems, and political campaign slogans. Flasks were also produced in a variety of shapes. Some resembled bellows or violins; others were globular or gourd-shaped.

Bitters bottles, those designed for various types of "patent medicines," are another important category. Most of these date to a time toward the end of the nineteenth century when the Women's Christian Temperance Union was raging against the evils of drink. Gin- and whiskey-sellers took to adding a few herbal ingredients to their basic product, labeling it "bitters," and selling it as a cure for a wide range of human disorders. These preparations were equal in alcoholic content to the Scotch whiskey or rye one buys today.

Bitters bottles ranged in color from light amber to dark brown. Usually they were square-shouldered. The bitters bottle was never without a label describing the wonders of what it contained.

The Pure Food and Drug Act of 1906 ended the bitters binge. However, during Prohibition patent medicines again enjoyed a spurt in popularity.

The western states usually are the best hunting grounds for whiskey and bitters bottles. The prospectors and miners of a century or so ago, were hardy individuals with a well-deserved reputation for their fondness for strong drink. Residents of old towns of the West usually buried their trash in pits, the bottles along with everything else. Sometimes they simply dumped trash into a creek. Never overlook a creek bed when searching through a western ghost town.

One way to establish the age of a bottle is by examining the mold mark, the upraised seam caused by the mold joint. Before 1860, the seam extended up the side of the bottle to a point just over the shoulder. On

Flasks of the nineteenth century.
*(Glass Container Manufacturers Institute)*

bottles produced between 1860 and 1880, the mold seam ran up the bottle's neck. Between 1880 and 1900, the seam extended the full length of the bottle, sometimes up and over the lip.

In the category of bottles, don't overlook the Mason jar, the wide-mouthed glass jar with the screw top, used widely during the nineteenth century for home canning and preserving. The Mason jar had a zinc screw-on cap, while its successor, the Ball jar, was equipped with a metal locking hinge. These fittings are what start your detector buzzing.

The Mason jar, invented by John Mason in 1858, was a notable step forward in the preservation of food. Until it came into existence, housewives had no practical method of sealing food against air. But the Mason jar, by providing an airtight seal, made home canning a common practice, and enabled families in the northern half of the United States to enjoy tomatoes, beans, peaches, and pears through the winters. Previously their winter fare had been limited to potatoes, squash, and turnips.

The first Mason jars were made of molded glass, were of quart size, and had a threaded top. Through the years, the design of the top changed several times, however, today's Mason jars are not very different from the earliest ones.

Mason jars you find may bear the name "Ball Mason's" in upraised lettering on the jar side. This is because Mason, in 1873, assigned the patent rights to Ball Bros. Glass Manufacturing Co. of Muncie, Indiana. By 1887, Ball Brothers was turning out 600,000 jars daily. The product came to be known as the Ball-Mason jar and, more recently, as, simply, the Ball jar.

The early jars had an aqua hue and were embossed with the phrase "Mason's Patent Nov. 30th 1858." These jars are worth $5 to $10. On the base of some of these appears "Pat. Nov. 26 67." Jars embossed "Mason's Improved" also bring $5 to $10. These have the same base lettering as mentioned above.

The bottles that you find aren't going to resemble those on display at the local antique shop, at least not in appearance. They'll be encrusted with dirt and mineral deposits and require a thorough cleaning before they begin to sparkle.

Start by giving the bottles a good soaking. Use a large metal tub if you're going to be cleaning several bottles at one time. Fill the tub with water and add ordinary household lye, following the directions on the package. Some people prefer to use ammonia. Let the bottles soak for several days. When you remove them, most of the encrusted dirt will fall away. Finish the job with a stiff brush.

Use a bottle brush to scrub the inside walls. If the brush is unable to reach all of the inside surface, partly fill the bottle with water, then add

Dump-sites of abandoned New England farms yielded these bottles for collector Aime LaMontagne. *(George Sullivan)*

fine sand—and shake. The abrasive action of the sand will act as a cleanser. Be sure it's *fine* sand; big grains or small pebbles may break the bottle. The final step is to rinse the bottle in warm water. Polish it with a soft cloth.

The subject of bottle collecting is enormous, and the paragraphs above cover some of the high points only. Fortunately, an excellent body of

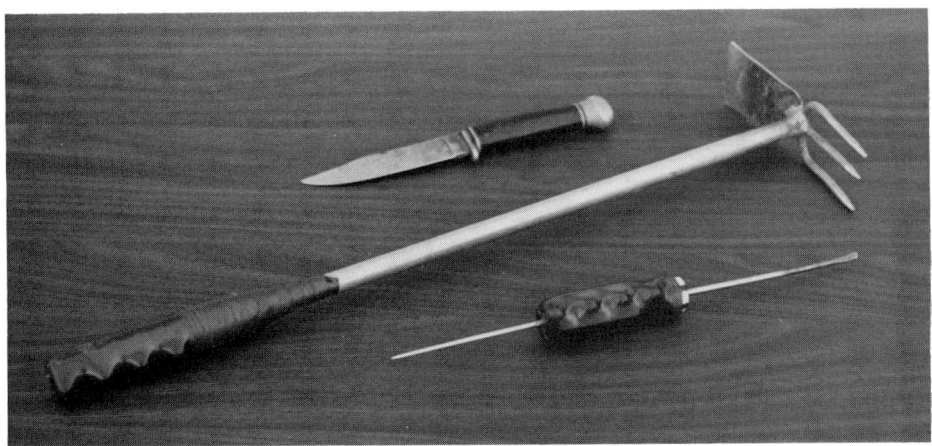

Common bottle-hunting tools. *(Garrett Electronics)*

literature is available on the topic, most of it produced in the past ten years. These are among the titles recommended:

> *The Collector's Book of Bottles,* by Marian Klamkin (Dodd, Mead & Co., 1971, $8.95), contains helpful information on American bottles of every type, even those dating to the earliest times. Every facet of the hobby is explored. There are more than 400 illustrations.
>
> *Poor Man's Guide to Bottle Collecting,* by Ferol Austen (Doubleday & Co., 1971, $7.95), is a fondly written, well-illustrated introduction to bottle identifications.

Glassmakers of a century or so ago would sometimes use the glass they had left over at the end of a day to make colored insulators for the telegraph or telephone companies. Collecting such insulators has become a hobby in its own right, and there are now clubs and organizations devoted to the speciality and several books have been written on the topic.

Insulators were used to insulate transmission wires from the pole crosspieces in the days before sheathed cables. Most stand 3½ to 4 inches in height and have a generally cylindrical or "petticoat" shape. Frequently they are pale green or aqua, but they also can be amber, olive green, or as clear as window glass.

Collectors learn to recognize different types by their flaring, domes, or grooves. "Skirts," "treads," and "drip points" are also used in distinguishing one type from another.

Like other glass objects, the older the insulator the greater its value. Insulators were first used on telegraph wires as early as 1844. Telephone companies began using them in the 1870s.

Embossed company names and dates are also important in establishing values. Most insulators range in price from 50¢ to $3, but a scarce specimen can bring as much as $40, and there have been a few instances of insulators being sold for as much as $100. Prices have moved steadily upward in the past decade, for the number of collectors is constantly increasing while the supply is fixed.

Where is the best place to look for insulators? Nearby overhead telephone lines, particularly at the base of telephone poles. When insulators were replaced, the repairman would often leave the "old" insulators strewn about the pole.

### RELICS OF IRON

Iron mining and working with iron were other early industries in colonial America. An ironworks was in operation in Virginia as early as

1621, and artifacts from a similar enterprise on the Saugus River near Lynn, Massachusetts, destroyed by the Indians in 1685, are still being recovered. Early forges and furnaces established in Pennsylvania on the Schuylkill River paved the way for America's giant steel industry.

There was no deep mining as we know it today. Colonists dredged up bog iron from along the streams or dug ore from shallow pits. Foreign artisans—German, English, and Scotch-Irish—were imported to work the iron. Sometimes they were assisted by bond servants, Indians, and even slaves. These iron workers turned out bar iron for forgings and pigs of more brittle iron which were used in casting.

The bars were worked and shaped—wrought—by a blacksmith with his anvil and hammer. The pigs were melted down to be poured and cast.

Ruins of old forges and furnaces have been found at countless sites throughout what was once colonial America. Still remaining are chunks of old pocked iron which was smelted over deep charcoal pits, fired by the trees from the countryside.

These early artisans produced many of the necessities of the day: nails, door latches and hinges, shovels and plowblades, wagon wheel rims and cooking utensils of every type. More specifically, the objects of old iron that collectors seek include the following:

*Fireplace Equipment*   Pokers, tongs, shovels, and andirons; trammels, long notched bars used for hanging cookpots at different heights; and S-hooks for holding the bails of pots and kettles.

*Cooking Utensils*   Pots and kettles; spiders, three-legged pans with short handles; trivets, three-legged stands made of iron for supporting cooking vessels in the fireplace; ladles and skimmers; spoons and forks.

*Nails*   Square-headed, hand-wrought nails are worth saving. There are countless others; one catalog lists 2,150 different types.

Early nineteenth-century door hardware. *(George Sullivan)*

Cast iron from a foundry of the eighteenth century. *(George Sullivan)*

*Lamps* Early lamps were known as grease, or slut lamps, and consisted of a saucerlike reservoir into which the wick was placed; sometimes these lamps had pivoted covers; Betty lamps, usually tin, were similar.

*Door and Shutter Hardware* Hinges, hasps, door latches, and locks; keys.

*Flatirons* The simple box iron had an opening in which a hot slug was placed; later flatirons had handles decorated with fancy curls and twisted forms; sadirons, larger and heavier, were pointed at both ends.

These categories represent only a sampling of what is to be found. There were candlestick holders and matchsafes, hitching posts and boot scrapers. Each of the various trades of the day had a whole array of tools: blacksmithing, harness-making, shoemaking, carpentry, wood turning, and wheel wrighting. Tools, by the way, can often be found in and around old barns.

Treasure seekers exploring abandoned homesteads and farm buildings in New England often come upon pieces of hand-wrought or cast iron of unusual shape and design, and can only guess at what the object might have been used for. Fortunately, the paraphernalia used by early Americans is a subject that has fascinated any number of people. As a result much has been written on the subject. Your local library is certain to have several informative and authoritative books. *The Complete Book of American Country Antiques,* by Katherine Morrison McClinton (Coward-McCann, Inc., 1967, $5.95), is a splendid introduction to the many items of wrought and cast iron, copper and brass used by colonial New Englanders and ranchers of the Old West.

## ON THE BATTLEFIELDS

Military relics are another flourishing specialty. Bullets and cartridge cases, cannon balls, uniform buttons and belt buckles, an occasional sword or pistol, bayonets, and random coins—these are the objects being recovered in abandoned forts and on American battlefields.

Proceed with caution, however. An Army private came upon live bullets and live grenades recently when searching a World War II battle site near Agana, Guam.

The continental United States is fairly well blanketed with former battlegrounds. At your local library or historical association, you can learn how your area figured in military activity of past.

One of the best source books on battlefields and where they are located is titled, *Alphabetical List of Battles, 1754-1900,* by Newton A. Strait. More than half of the book's 250 pages are given over to a careful and

Searching for military relics at a World War I battery site. *(George Sullivan)*

detailed listing of Civil War military engagements, along with a chronological history of the War. But the author also lists battlefields of the American Revolution, the War of 1812 (when such towns as Fort Meigs, Ohio, Fort Harrison, Indiana, and Frenchtown, Michigan, were the scene of armed fighting), and the French and Indian War (with Oswego and Lake George, New York, and Monongahela, Pennsylvania, among the battle scenes). The author doesn't overlook the Mexican War or what he terms the Florida or Seminole War.

Compiled from official records, the book was published in Washington, DC, in 1906. It is likely that your library has a copy. Fortunately, the book was republished in 1968 by Gale Research (The Book Tower, Detroit, MI 48226) and is available by mail.

You can also track down American battlefields by consulting good general histories of particular wars. *The Civil War, Day by Day, an Almanac,* by E. B. Long (Doubleday & Co., 1971) and *The Encyclopedia of the American Revolution,* by Mark Mayo Boatner, III (David McKay Co., 1966) are two such histories.

## RINGS AND OTHER JEWELRY

Speak to anyone who has been coinshooting for a year or so, and he'll tell you of the many rings and other items of jewelry that he's found. It is a fact that you will recover about three or four rings for every hundred coins you find.

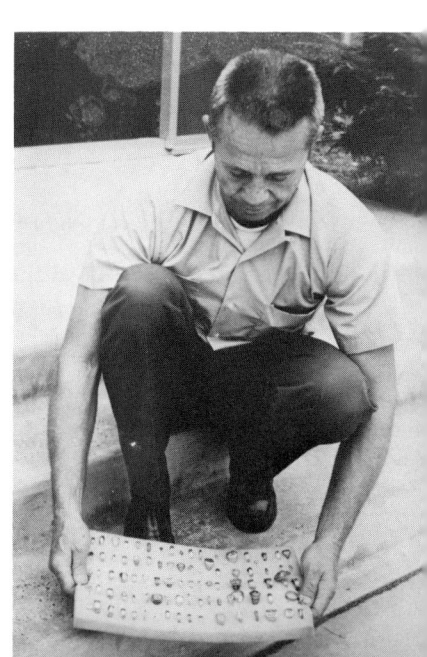

If you do much coinshooting, you're sure to find rings. *(The Goldak Company)*

Rings are made of alloys of various types and often the metal molecules, like those of copper, leach into the surrounding soil, creating a detectable area larger than the ring itself. Thus, treasure hunters are constantly reporting the finding of rings that have been lost for years. In 1974, a Dublin, California, man found a two-diamond ring that his daughter had lost while practicing baton twirling in the backyard eleven years before.

It used to be fairly standard practice to polish recovered rings and other jewelry with the idea of selling them. But with the price of gold and silver skyrocketing, some rings are now frequently sold for their precious metals value. A man's gold wedding band that sold for $20 or $30 in 1971 was being retailed for $50 or $60 in 1974.

## BUTTONS

No matter where you do your treasure hunting, you're almost certain to find buttons, metal buttons. Historical sources say that buttons were being made in America as early as 1706.

More buttons have been made of brass than any other metal or alloy. Brass has always been available; it is easy to work. Steel as a button material dates to the mid-eighteenth century. At first, steel was considered a luxury metal, its value equated with that of gold and silver. Pewter, a tin alloy, was the poor man's button material for centuries. Aluminum buttons are of fairly recent vintage, the first ones dating to the early 1900s.

Military buttons form one of the most popular collecting categories. The number of different American buttons, solely military in nature, is well up in the hundreds. Other popular occupational buttons are those from the uniforms of policemen, firefighters, railroad men, and streetcar conductors. Historical and political buttons are also highly prized.

These paragraphs are only meant to suggest the infinite number of button types and collecting categories. It is a subject that is rich enough to fill many books.

You can get some idea of the diversity of the subject from a book titled, *The Button Sampler,* by Lillian Smith Albert and June Ford Adams (Gramercy Publishing Co., 1961, $1). The National Button Society of America (Box 116, Lamoni, Iowa 50550) can prove helpful, too.

## HUNTING FOR PRIZES

One outgrowth of the surging interest in the use of metal detectors is the organized treasure hunt. Common to many Midwest locations, a treasure hunt usually consists of a free-for-all search over a carefully

# $10,000 TREASURE HUNT

Sponsored by the Central Okla. TH'ers Club & Shepherd Mall Merchants Assoc.

**$1,000 CASH** ➤ FIRST PRIZE IN THE COMPETITION HUNT WITH CASH PRIZES TO 2, 3, 4th!

PLUS

**$5,555.55** ➤ CASH IN SILVER AND GOLD COIN... PLUS METAL DETECTORS AND OTHER VALUABLE PRIZES!

## WHERE?
### Oklahoma City Fairgrounds
Oklahoma City, Oklahoma

## WHEN?
### APRIL 21st, 1974

LIMIT OF 600 PARTICIPANTS—MAIL YOUR ENTRY FEE NOW! $20.00 COVERS YOUR COST TO BOTH COMPETITION AND TREASURE HUNT.

---
Mail to: Shepherd Mall Post Office, Oklahoma City, Okla. 73107

NAME ..................................................
ADDRESS ..............................................
CITY ........................ STATE ......... ZIP .........

Rain or shine no refunds. Send check or money order!
---

For additional information write: Bob Barnes, 2720 Villa Prom, Suite 200, Oklahoma City, Okla. 73107

## SEE APRIL 19-20 TREASURE SHOW!

Shepherd Mall Merchants Assoc. have changed the November show back into April—see and meet the greats Karl Von Mueller, Bill Mahan, Ed Bartholomew, Charles Garrett and many others. See exhibits from all over the world Fri. and Sat., April 19 and 20 in 1974.

# shepherd mall
VILLA TO PENN. ON N.W. 23rd ● Okla. City, Okla.

defined area that has been previously seeded with coins and other finds. The total amount of prizes offered can be substantial. At the Midwest Treasure Finders Hunt, held annually in Plainfield, Indiana, over $5,000 in prizes are awarded. Usually there are separate contests for men, women, and children.

Your entry fee subsidizes the cost of most of the prizes. Sometimes local merchants donate merchandise for prizes in exchange for the promotion and good will they receive. Numbered metal tokens are then seeded in the search area, each corresponding to a piece of merchandise. Sometimes manufacturers of metal detectors donate one or more of their models to be used as prizes.

Many such treasure hunts have overtones of a sales convention. Equipment manufacturers and their dealers are on hand, displaying the latest in detectors and accessory equipment. If the hunt is sponsored by a particular club, members often display objects that they've recovered over years. You're likely to see exhibits of rare coins, buttons, bottles, military artifacts—any discovery a metal detector might have been responsible for.

If you're a treasure-hunting novice, competing in such a hunt is recommended, not merely for the prizes you might "win" but more for the opportunity you'll have to meet experienced treasure seekers and listen to their tales. You're sure to pick up plenty of valuable pointers. The equipment dealers on hand will also give you tips on the use of detectors.

Where do you find out about such meets? A local treasure-hunters club is sure to know of them. The bigger events are advertised in the pages of *True Treasure* and *Treasure World.* The Prospectors Club International (P.O. Box 68069, New Augusta, IN 46268) and the National Treasure Hunters League (Box 53, Mesquite, Texas 75149) sponsor and promote several hunts each year.

When entering such contests, use a digging tool that enables you to get the job done quickly, a short-handled hoe-type instrument or a small three-pronged rake. Here, neatness counts for nothing.

OTHER DETECTOR USES

During the summer, a New England treasure hunter keeps an eye on the trout streams in his area, and makes a mental note of which ones are the most popular and what spots are the anglers' favorites. Then in the fall he searches underwater along the banks with his detector looking for snagged fishing lures. A good day of "fishing" nets him a dozen or so. Some are worth $4-5 apiece.

The owner of a deep-seeking detector, one capable of perceiving large objects at several feet, happened to have been employed in the oil business at one time, and he knew the value of the diamond-tipped drill bits that companies use in drilling new wells. He also knew that when wells come in, they frequently explode mud and debris over everything in sight. Drill bits often get buried and are never recovered. So he started searching the abandoned drilling sites. Before long, he had recovered seventeen of the expensive bits, each of which earned him a cash reward from the drilling company.

An Indianapolis treasure hunter uses his detector to search for copper wire discarded by the work crews of local telephone and power companies. He does his searching wherever power lines have been erected, following them over hill and dale for miles at a stretch. He takes the recovered wire home, puts it in a metal barrel, sets it afire to remove the insulation, then sells the copper that remains to a local scrap-metals dealer.

Another enterprising detector owner uses his machine to recover brass shells by the thousands from local outdoor firing ranges. He sells them to a local gun shop and they reload them. Cartridge cases, being brass, also have value in the scrap-metals market. So do keys, which are also made of brass.

These are only some of the methods used by detector owners to produce income. Many keep an eye on lost-and-found advertisements in the local newspaper. When they read that someone has lost a valuable object—a watch, ring, important keys, or whatever—they call the person and offer their services, usually arranging an hourly rate in advance.

Some treasure hunters have even established part-time businesses as specialists in the recovery of lost objects, lost *metal* objects, that is. The first step in getting started in such an enterprise is to have business cards printed bearing your name, address, and telephone number, together with a slogan, something like, "I find lost valuables: rings, watches, rare coins, etc." Then post the cards on bulletin boards in supermarkets, laundromats, and especially at resorts or beach clubs. Somebody is always losing something at such places.

If your equipment is capable of being used underwater, distribute the cards at marine supply shops, boathouses, and tackle shops. During one summer, one freelancer was called on several times to find outboard motors that had ended up in the water. He was successful on each occasion. Another man was able to restore harmony to a couple's marriage by finding the wife's diamond engagement ring which she had lost in a snowdrift while unloading packages from the family automobile.

## TREASURE HUNTING AND TAXES

If you're successful as a treasure hunter, the subject of income taxes may be of concern to you. It's not all bad, however. Should you make a declaration of income gained through treasure hunting, you are also likely to be entitled to list certain expenses, i.e., deductions.

Suppose you're a coinshooter and in the course of a calendar year you recover coins with a face value of $215, say. Must that amount be declared as income? It depends. If you put the coins in glass jars and store them in a closet, hoping that they will appreciate in value, then you don't have to declare them as income. But if what you find simply gets mixed in with your pocket change, it's a different matter. This money must be declared.

Generally speaking, the same holds true for all the other items you find—bottles, iron relics, jewelry, etc. If you have a bottle collection that you believe is worth several hundred dollars, and you keep it on display, it's not regarded as income. But sell the bottles—or sell only one bottle—and the amount received becomes income.

What about expenses? Once you've begun to derive income from treasure hunting, you are likely to be entitled to deduct some of the costs related to the activity. The amount you paid for the metal detector you use and the various accessories fall into the category of expenses.

That's not all. Suppose you're a part-time prospector, spending fifteen or twenty weekends a year up in the foothills with a surface dredge and various other paraphernalia. Besides your equipment costs, you also can deduct what you pay for gasoline going to and from the gold-digging site. You can deduct the cost of lodging. However, such expenses cannot exceed the income you derive.

If, when you file your income tax, you do plan to declare income from treasure hunting and list the expenses you incurred, use Schedule C (of Form 1040). It is titled "Profit (or Loss) for Business or Profession."

One other matter. If you make a particularly important find one year—an object worth several thousand dollars, a gold nugget, say—it may be worthwhile for you to spread out the money you receive when you sell it over several years. "Income averaging," this is called. Whether you will be entitled to estimate your tax on this basis depends on the amount netted from the find as well as your previous taxable income in past years. The I.R.S. used this formula in 1974: "If after subtracting $3,000 from your 1973 taxable income, the balance is over 30 percent of the total of your taxable income for the last four years, 1969 through 1972," then income averaging is permitted. You use schedule G (of Form 1040) when calculating your tax on this basis. For more information, obtain publication 506 from your local I.R.S. office. It is titled, *Computing Your Tax Under the Income Averaging Method.*

Keep in mind that federal tax regulations change from year to year, and so do interpretations of the rules. It's wise to consult an I.R.S. representative or a tax accountant for current advice on the topics covered here.

Chapter 6

# ALL ABOUT GOLD AND PROSPECTING

Gold! On weekends they trek by the thousands into the Mother Lode country of California, driving up Route 49 in their vans and campers, hauling their dredges, gold pans, and diving gear. Prospecting and small-scale mining may not yet rival pro football-watching or even tennis as the nation's favorite recreational activities, but their growth in recent years has been enormous.

California isn't the only state affected. Panning clinics are being offered in Oregon. Equipment dealers in Colorado are hard-pressed to keep metal detectors and portable mining gear in stock. And in Montana, the number of mining permits issued jumped tenfold in one recent year.

The reason for all of this was summed up recently in a headline in the *New York Times:* "SOARING PRICES BRING NEW GOLD RUSH TO THE WEST." Soaring is the appropriate word.

The price of free market gold, $35 an ounce in 1970, zoomed to $65 an ounce in 1972. In May the next year, it broke the $100 barrier and kept right on going. In 1974, in February, the price touched $184 an ounce. And all the while the value of the dollar continued to sag.

Some recent history has a direct bearing on what's been happening today. Although the United States abandoned the gold standard more than forty years ago, a strong tie remains between gold and the dollar. At the Bretton Woods conference after World War II, the countries that established the International Monetary Fund agreed that gold would continue to be used for settling international debts and that the dollar would be the key currency for organization members.

At first, the nine nations that made up the IMF gold pool kept the price of gold on the London market from rising above $35 by the simple expedient of selling off their holdings whenever there was buying pressure. But during the monetary crisis of 1968, triggered by the weakness of the dollar, the gold-pool nations switched policy, agreeing not to sell any more gold. Since that time, there have been two prices for gold. There is the official price used in dealings between nations. Late in 1973, the official price was $42.22. Then there's the free market price, the one that keeps acting like a thermometer on an August day.

Many of today's gold seekers prefer to stash away the small vials of tiny gold flakes and particles they recover. (While a private citizen cannot own gold bullion, he is permitted to have as much naturally occurring gold as he can find.) Selling it, however, is no problem. Jewelers, dental suppliers, and industrial concerns need gold, six to eight million ounces a year, and they have to pay the free market price to get it. The Yellow Pages of the San Francisco telephone directory list a half a dozen dealers in gold and silver.

Can the average person find enough gold to reimburse him for his time and his equipment costs? Indeed, he can, provided he makes a careful study of the mining record and geology of the area in which he's interested and then employs sophisticated techniques in prospecting and mining. Says an official of the U.S. Geological Survey: "The development of new highly sensitive and relatively economical methods of detecting

In some areas, gold panning is a family enterprise. *(State of Montana, Department of Highways)*

gold has greatly increased the possibility of discovering gold deposits too low-grade to have been recognized heretofore by the prospector armed only with a gold pan...."

To sum up, the chances of making an important find, a "strike," are terribly slim, fewer than one in a thousand. The chances of boosting your income a little are very good. If you're just out for a good time, as many gold seekers are, well, the odds are all in your favor.

### WHERE TO LOOK

When you think of gold and prospecting, you conjure up scenes of the Old West, of bearded prospectors and balky burros and the rugged terrain of the Sierra Nevada, or desert wastelands. The truth is that gold has been found in just about every one of the fifty states. It has been mined commercially not only in California, Colorado, and Nevada but also in Georgia, Pennsylvania, North Carolina, and Massachusetts.

Of course, your chances of finding gold are the best if you do your prospecting on the Pacific Coast or in the Rocky Mountain West. Some parts of California are virtually awash with gold-bearing streams and rivers. Out of the mountains north of Los Angeles flow the rivers Kern, Placerita, and San Gabriel, all well known for their placer deposits. The same is true in the northwest corner of the state where you have the Klamath, Scott, Salmon, Trinity, Shasta, McCloud, Pit, and Hay Fork rivers from which to choose.

That's only part of it. A strip of land approximately 50 miles wide and 125 miles long lies along the western slope of the Sierra Nevada range which is known as the Mother Lode, once one of the richest gold regions in the world. The Mother Lode is not, as many people think, one long vein, but a series of small parallel vein systems. The richest part of the network covers an area about one mile wide extending from Mariposa in the south to Georgetown in the north.

A great many rivers cut westward through the region to converge east of San Francisco then flow into San Pablo Bay. The Sacramento River courses in from the north, fed by the Butte, Feather, Yuba, Rubicon, American, and many smaller streams. The San Joaquin River flows in from the south and is joined by the Stanislaus, Tuolumne, Merced, Chowchilla, and Kings rivers. The Consumnes and Moklumne rivers flow into the bay directly out of the east.

There is also enormous mineral wealth in Colorado. The Cripple Creek goldfield just west of Colorado Springs was discovered in 1891. By the turn of the century it was producing almost $20 million annually. The

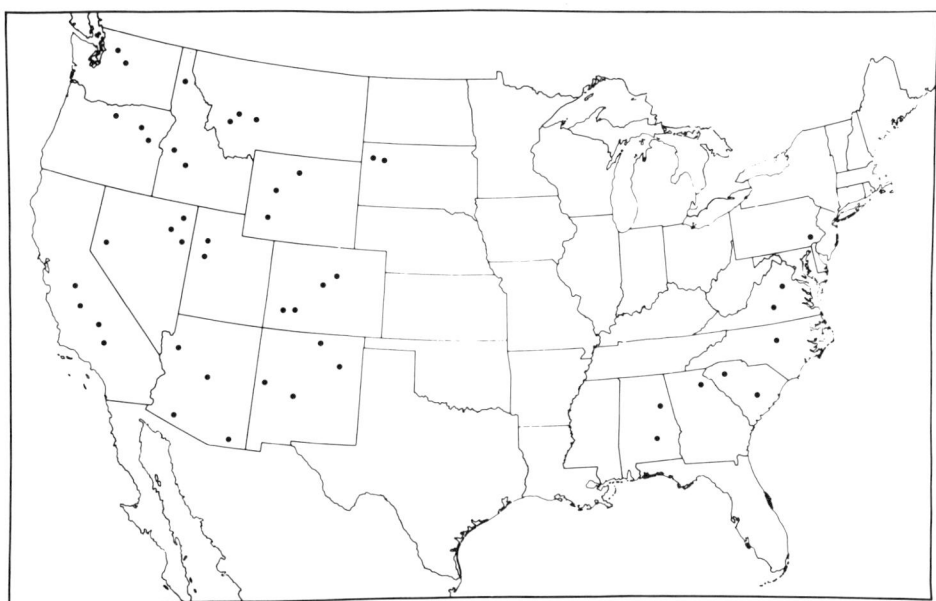

Major gold producing areas in the United States. *(U.S. Department of the Interior, Geological Survey)*

area is the center of frantic activity today. Gold bearing rivers in Colorado include the Arkansas, San Juan, upper Rio Grande, Cherry Creek, South Platte, Eagle, and the upper Colorado. Virtually every stream in Boulder, Eagle, Clear Creek, Park, Teller, and Lake countries near Denver carries gold particles, as do the rivers of San Miguel and San Juan counties in the southwestern part of the state.

Colorado gold that washes into the San Juan and Colorado rivers is also recovered in southeastern Utah. Rivers and streams in southwest Montana carry gold from the Yellowstone region.

The rivers of western Nevada, the wide strip of land between the borders of Oregon and California, are known to be rich in gold deposits, particularly those in the Virginia City-Carson City areas. These include the Reese and Humboldt rivers, the Walker River, both its east and west tributaries, and the Carson River. The Black Hills of South Dakota have yielded important quantities of gold in the past and hold promise for the future. They're traversed by Fourche and Cheyenne rivers, which, along with their tributaries, often "show color," a term meaning they display tiny flakes of gold.

In the state of Washington, the Columbia River contains gold particles brought from the Cariboo Mountains of British Columbia. In Oregon, the Rogue and Bear rivers and their tributaries are the ones that prospectors focus upon. The Snake River in Idaho brings gold from the Yellowstone region, while the Kootenai River, located at the state's northern tip, carries gold in from Canada.

In Montana, placer gold has been found in the southwestern part of the state, chiefly within the Missouri River and its tributaries. Other important districts are in Madison County, which ranks first among counties in placer gold deposits (and third in total gold production).

One of the principal dredging areas for placer gold in Idaho is centered in the Boise Basin of the Middle Fork of the Boise River just a few miles northeast of Boise itself. Other well-known placer areas are to be found along the Salmon River in Lemhi and Idaho counties, and on the Clearwater River at its tributaries, particularly at Elk City and Orofino.

Books on gold prospecting are so popular in Oregon that librarians have difficulty keeping them on the shelves. In terms of placer gold, interest is centered on the tributaries of the Rogue River and the neighboring streams on the Klamath Mountains. The main streams that drain the Blue and Wallowa mountains are also the subject of placer mining. As for lode deposits and the mining activity associated with them, the most important regions are in the northeastern part of the state.

In South Dakota, small amounts of placer gold have been produced in the Black Hills, particularly in the Deadwood region, on French Creek (near Custer), and in Washington on the Columbia and Snake Rivers.

As for the eastern states, limited amounts of gold have been washed from some of the streams that drain the eastern slope of the southern Appalachian Mountains, including parts of Maryland, Virginia, North Carolina, South Carolina, Georgia, and Alabama. A small amount of gold has been recovered in the New England states. Gold panners are to be seen occasionally in Maine along the banks of the Swift River, and in New Hampshire operating in Indian Stream, which is found at the northernmost tip of the state. "Other placer deposits in the East may be discovered," says the U.S. Geological Survey, "but they will be of low grade, costly to explore, and difficult to recognize."

The paragraphs above hit only the high spots. You have at least some chance of finding gold in these areas of the United States:

| State | County |
| --- | --- |
| Alabama | Chilton, Clay, Cleburne, Colbert, Randolph, Talladega, Tallapoosa. |
| Alaska | (Alaska does not have county designations.) Regions where gold is found include Prince William Sound, Kenai Peninsula, eastern part of Kushkokwin Valley, Yukon Basin, Seward Peninsula, Kobuk River. |

| State | County |
|---|---|
| Arizona | Apache, Coconino, Gila, Graham, Maricopa, Mojave, Pima Pinal, Santa Cruz, Yavapai, Yuma. |
| Arkansas | Garland, Saline. |
| California | Amador, Butte, Calaveras, Colusa, Contra Costa, Eldorado, Fresno, Imperial, Inyo, Los Angeles, Madera, Manposa, Mendocino, Mono, Napa, Nevada, Placer, San Bernardino, San Diego, Siskiyou, Trinity, Tuolumne, Ventura. |
| Colorado | Adams, Alamosa, Arapahoe, Baca, Boulder, Chaffee, Clear Creek, Conejos, Costilla, Custer, Denver, Dolores, Douglas, Eagle, Elbert, El Paso, Fremont, Garfield, Gilpin, Grand, Gunnison, Hinsdale, Huertano, Jackson, Jefferson, Lake La Plata, Mineral, Moffat, Montezuma, Montrose, Ouray, Park, Pitkin, Rio Grande, Routt, Saguache, San Juan, San Miguel, Summit. |
| Georgia | Cherokee, Forsyth, Gordon, Hall, Lumpkin, McDuffie, Paulding, White, Wilkes. |
| Idaho | Adams, Blaine, Boise, Boundary, Cassia, Clearwater, Custer, Elmore, Gem, Idaho, Kootenai, Lemhi, Owyhee, Shoshone, Valley. |
| Illinois | Hardin. |
| Indiana | Brown, Franklin, Jennings, Monroe, Morgan, Northington, Warren. |
| Iowa | Jackson. |
| Kansas | Ellis, Gove, Trego. |
| Maine | Cumberland, Franklin, Hancock, Knox, Oxford, Penobscot, Somerset, Waldo, Washington. |
| Maryland | Baltimore, Montgomery, Prince George. |
| Massachusetts | Essex, Hampden, Hampshire, Worcester. |
| Michigan | Marquette. |
| Minnesota | Benton, Fillmore, Itasca, Kandiyohi, Olmsted, St. Louis, Wabasha. |
| Mississippi | Jackson. |
| Missouri | Adair, Macon. |
| Montana | Beaverhead, Broadwater, Cascade, Chouteau, Deer Lodge, Fergus, Granite, Jefferson, Lewis and Clark, Madison, Mineral, Missoula, Park, Phillips, Powell, Ravalli, Silver Bow. |
| Nebraska | Franklin, Harlan, Seward, Stanton. |
| Nevada | Churchill, Clark, Douglas, Elko, Esmeralda, Eureka, Humboldt, Lander, Lincoln, Lyon, Mineral, Nye, Ormsby, Pershing, Storey, Washoe, White Pine. |
| New Hampshire | Carroll, Coos, Grafton. |
| New Mexico | Colfax, Grant, Lincoln, Rio Arriba, Sandoval, Santa Fe, Sierra, Socorro. |
| New York | Allegheny, Dutchess, Erie, Fulton, Hamilton, Herkimer, Rockland, Saratoga, Washington, Westchester. |

| State | County |
|---|---|
| North Carolina | Burke, Cabarrus, Caldwell, Catawba, Cherokee, Clay, Cleveland, Davidson, Franklin, Gaston, Granville, Guilford, Henderson, Jackson, Lincoln, McDowell, Mecklenburg, Moore, Montgomery, Nash, Person, Polk, Randolph, Rowan, Rutherford, Stanley, Transylvania, Union, Watauga, Wilkes, Yadkin. |
| Oregon | Baker, Clackamas, Coos, Crook, Curry, Douglas, Grant, Harney, Jackson, Josephine, Lane, Lincoln, Malheur, Marion, Wheeler. |
| Pennsylvania | Chester, Lebanon. |
| South Carolina | Abbeville, Cherokee, Chesterfield, Edgefield, Greenville, Kershaw, Lancaster, Laurens, Oconee, Pickens, Saluda, Union, York. |
| South Dakota | Custer, Lawrence, Pennington |
| Tennessee | Blount, Monroe, Polk. |
| Texas | Brewster, El Paso. |
| Utah | Beaver, Box Elder, Grand, Iron, Juab, Kane, Piute, Salt Lake, San Juan, Tooele, Uinta, Utah, Wasatch. |
| Vermont | Bennington, Windsor. |
| Virginia | Buckingham, Culpeper, Floyd, Fluvanna, Goochland, Montgomery, Orange, Prince William, Spotsylvania, Stafford. |
| Washington | Asotin, Chelan, Clallam, Clark, Douglas, Ferry, King, Kittitas, Lincoln, Okanogan, Snohomish, Stevens, Thurston, Whatcom, Yakima. |
| Wisconsin | Clark, Douglas, Polk. |
| Wyoming | Big Horn, Crook, Fremont, Hot Springs, Johnson, Laramie, Sheridan. |

Whether prospecting East or West, topographic and geologic maps are vital. Topographic maps are those that present a precise representation of the physical configuration of an area. One way in which they do this is by means of contour lines, lines which join points of equal elevation and, thus, define surface configurations. Topographic maps covering every part of the country, scores of different ones for each state, and prepared in marvelous detail, are available through the U.S. Geological Survey (Branch of Distribution, 200 South Eads St., Arlington, VA 22202). Map stores often sell such maps, too. Chapter Four discusses these maps in detail. Maps of the national forests can be obtained from the appropriate forest supervisor or ranger, or by writing to the Forest Service (U.S. Department of Agriculture, Washington, DC 20250).

PROSPECTING ON FEDERAL LANDS

For the amateur prospector, perhaps the greatest opportunity for success lies within the federal government's so-called national resources

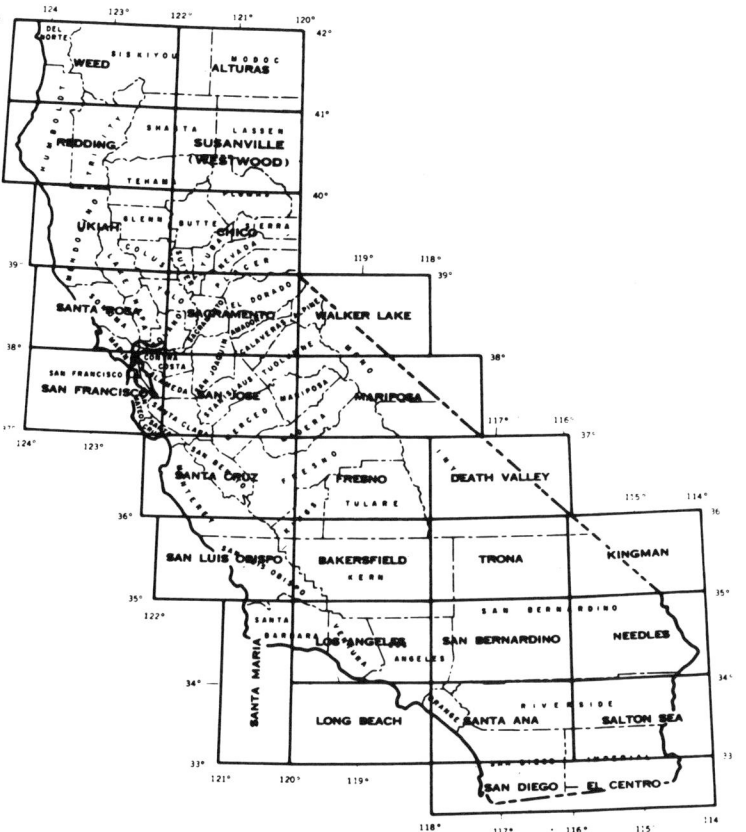 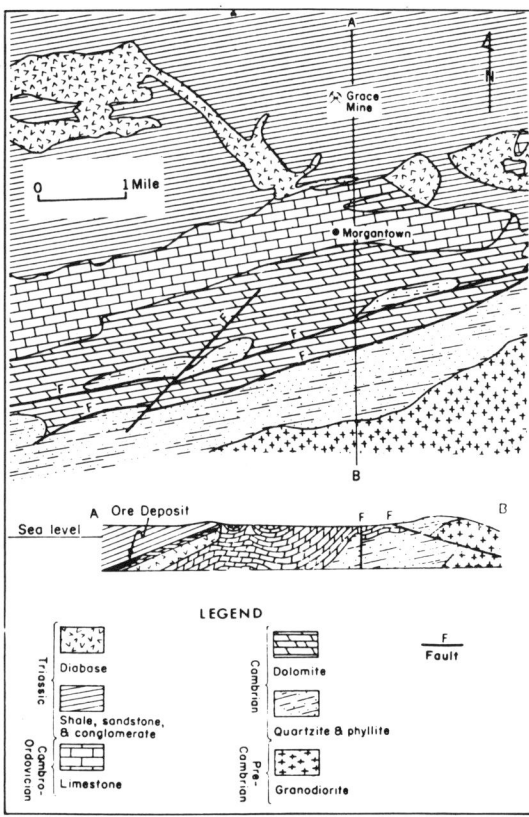

Topographic maps of California are available in "quadrangles" shown here. *(U.S. Department of the Interior, Geological Survey)*

A generalized geological map of the Morgantown area, Berks County, Pennsylvania. *(Pennsylvania Topographic & Geological Survey)*

lands, formerly known as public lands. These are vast regions under federal ownership which have not yet been set aside for national forests or national parks. They cover about one-fifth of the land area of the United States, more acreage than all the other federal lands combined.

Administered by the Bureau of Land Management, most of these lands are in the West and in Alaska; a few small tracts are in the Midwest and South.

In the administration of these lands, a "multiple use principle" is followed. This means that such activities as recreation, timber production, grazing, and mining are encouraged. In fact, the only places where you can't prospect and mine are those relatively small areas set aside as protective withdrawal sites, which have been so designated to preserve some unique natural resource. You are permitted to use whatever hand tools are necessary to your search, and metal detectors are allowed. In some areas, a permit is required before you may use heavy machinery, a surface dredge (see below), for example, or a recreational vehicle.

It is always wise to contact the appropriate office of the Bureau of Land Management to determine the rules and regulations in effect. Sometimes special authorization is required. In addition to information on regulations, some offices are extremely cooperative in providing maps and detailed advice as to an area's geological resources. Here is a listing of the regional offices and their addresses:

*Alaska*

Anchorage Land Office
555 Cordova St.
Anchorage, AK 99501

Fairbanks District & Land Office
516 Second Ave.
Fairbanks, AK 99701

*Arizona*

Arizona Land Office
Federal Building
Phoenix, AZ 85025

*California*

Riverside District & Land Office
1414 Eighth St.
Riverside, CA 92502

Sacramento Land Office
Federal Building
650 Capitol Mall
Sacramento, CA 95814

*Colorado*

Colorado Land Office
Federal Building
1961 Stout St.
Denver, CO 80202

*Idaho*

Idaho Land Office
Federal Building
Boise, ID 83701

*Montana (North Dakota, South Dakota, Minnesota)*

Montana Land Office
316 North 26th St.
Billings, MT 59101

*Nevada*

Nevada Land Office
Federal Building
300 Booth St.
Reno, NV 89502

*New Mexico (Oklahoma)*

New Mexico Land Office
Federal Building
South Federal Place
Santa Fe, NM 87501

*Oregon and Washington*

Oregon Land Office
729 Northeast Oregon St.
Portland, OR 97232

*Utah*

Utah Land Office
Federal Building
Salt Lake City, UT 84111

*Wyoming (Nebraska, Kansas)*

Wyoming Land Office
U.S. Post Office
2120 Capitol Ave.
Cheyenne, WY 82001

*Eastern United States*
  *(All Other States)*

Eastern States Land Office
7981 Eastern Ave.
Silver Spring, MD 20910

You are also permitted to prospect on most lands under the jurisdiction of the National Forest Service. There are 154 national forests and they cover an enormous area, 181 million acres, about one acre for every United States citizen. There's a national forest within a day's automobile trip of any point in the United States (except Hawaii and Alaska's deep interior).

When planning to prospect on national forest land, first check with the district ranger or supervisor's office. One or the other will advise you whether any restrictions apply to the land you plan to search. These sources are also helpful in supplying maps and other information. The Regional Offices of the Forest Service provide similar material. Here is a list of them, together with addresses:

Northern Region
Federal Building
Missoula, MT 59801

Rocky Mountain Region
Federal Center
Building 85
Denver, CO 80225

Southwestern Region
517 Gold Ave., SW
Albuquerque, NM 87101

Intermountain Region
324 25th St.
Ogden, UT 84401

California Region
630 Sansome St.
San Francisco, CA 94111

Pacific Northwest Region
319 SW Pine St.
P.O. Box 3623
Portland, OR 97208

Eastern Region
633 West Wisconsin Ave.
Milwaukee, WI 53203

Southern Region
1720 Peachtree Rd., NW
Atlanta, GA 30309

Alaska Region
Federal Office Building
P.O. Box 1628
Juneau, AK 99801

## PLACER GOLD

The great majority of amateur prospectors concentrate their energies on alluvial deposits of gold, which are known as placers. Mountain streams and rivers, cutting through terrain where gold-bearing veins are to be found, carry with them tiny flakes of gold and minute particles which have been washed from the veins through weathering and erosion. It is much easier to find these concentrations than the veins themselves.

Any place where the velocity of the water slackens is a likely spot to look for placer gold; for example, wherever the stream widens, where there's a reduction in the stream's gradient, or where a small stream enters a larger one. In such places, the gold particles get a chance to settle, to accumulate.

The downstream side of bar rocks that lie perpendicular to the stream's flow, and the upstream faces of large boulders also offer potential. Don't overlook riffles in bedrock, or cracks or potholes in stream boulders.

As for the different types of placer gold to be found, the Colorado Bureau of Mines gives this information:

Coarser gold is deposited in the upper part of the stream; finer gold in the lower portions.

The richer and coarser gold is deposited in the layers of comparatively coarse gravel and wash; finer gold is deposited in the finer sandy drifts.

The best gold occurs in the layers of wash containing black sand and pebbles of magnetite or other heavy material.

Placer gold doesn't always have the traditional appearance of gold. It can vary in color from a deep reddish yellow to a pale silvery white. No matter what it looks like, certain characteristics are evident. Pure gold has a specific gravity of 19.3 (which means that a given mass of gold is 19.3 times heavier than an equal volume of distilled water); that of placer gold is almost as high, ranging upward from 17.19. Lead, by comparison, has a specific gravity of 11.35.

In contrast to placer deposits is lode gold, deposits of gold within solid rock. Most areas of the country that are likely to contain lode deposits have been thoroughly explored. "The inexperienced prospector," says the U.S. Geological Survey, "without ample capital has very little chance of discovering a lode rich enough to be developed."

## FINDING PLACER DEPOSITS

As anyone who has ever washed out a pan of sand and gravel will tell you, gold particles are often to be found amidst concentrations of magnetic black sand. It's not that the sand attracts the gold by virtue of its magnetic properties. The flecks of gold and the grains of black sand, being of somewhat the same specific gravity, tend to accumulate in the same places.

You can take advantage of this fact by using a metal detector to locate black sand concentrations on the bottom of a creek or river. Once found, the sand can be sucked up with a small dredge, or shoveled up, then sluiced and panned.

A BFO-type detector is recommended when searching for black sand pockets. Tune it so you get a moderate beat.

Modern gold prospecting. *(Compass Electronics)*

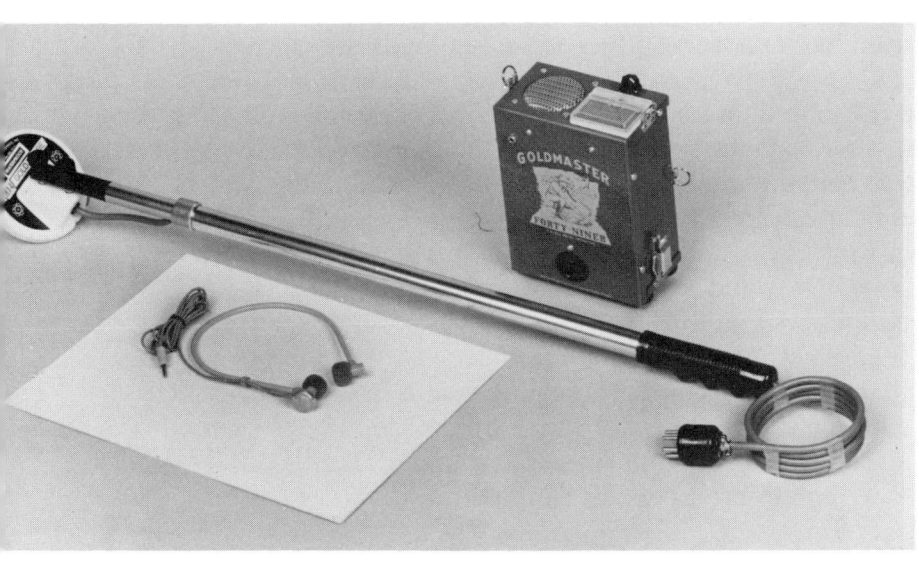

The Gold Master Forty-Niner; probe is fitted with a 6-inch coil. *(White's Electronics)*

The Alaskan Gold Master is a sophisticated TR system. *(White's Electronics)*

The size of the search coil should be determined by the conditions you're going to encounter. If the bottom of the stream is rough and strewn with rocks, use a smaller coil, one 4 to 8 inches in size. You'll be searching between rocks and boulders and you won't need much depth. When working a smooth-bottomed stream that's mostly sand, you'll need to get all the depth you can, so use an 8- to 12-inch coil.

It is usually best to tune the instrument to the metal mode when using the smaller coil. When the search loop is passed over concentrations of black magnetic sand, the beat rate will moderate, slow down. But should the loop be passed over a gold nugget, a metallic substance, that is, the beat rate will increase.

When using the larger search loop, tune in the mineral mode of operation. The 8- to 12-inch coil isn't going to detect small metallic objects—like nuggets—anyway. The mineral setting assures optimum sensitivity in tracking down black sand deposits. The beat rate will increase any time the search loop passes over a sand pocket.

It's important to bear in mind that the water temperature of a cold mountain stream has an adverse affect on some brands of detectors, signal drift especially. Be sure that the detector you plan to use is drift-free.

When working mountain streams, the detector's control box becomes a "body mount" unit; it's either strung around the user's neck or fixed to his belt. The search coil, mounted to a 4-foot shaft, is linked to the control box by means of a waterproof flexible cable that is several feet long.

This system enables the operator to move about in the stream with the greatest possible facility and reach easily with the probe in every direction. It also allows him to explore water that's several feet deep without any fear of getting the control box wet.

For finding accumulations of black magnetic sand in stream beds (wet or dry), Garrett Electronics offers a 3½-inch waterproof loop to be used with the standard Garrett BFO machine (see Chapter 1). This coil is especially meant to detect sand deposits in bedrock cracks or fissures. Each is priced at $25.95.

For those who prefer a TR machine, White's Electronics has two units especially designed for prospecting. There is the Gold Master Forty-Niner, in which the control case straps to the user's chest. The probe he carries can be equipped with either 4- or 6-inch waterproof coil. The Forty-Niner is $149.50.

White's Alaskan Gold Master (Model 70-TR) is a more sophisticated machine, equipped with an intensity meter, mineral/metal control knob, battery check, and earphone jack (none of which is included with the Forty-Niner). A 4- or 6-inch coil comes with the unit. The Alaskan sells for $269.50.

Also for TR adherents, there's the Compass equipment line. The company's Yukon 76-B can be adapted as a body-mount unit.

## PANNING

In the gold country of northern California, that region extending from the foothills below Yosemite north almost 200 miles into the beautiful valleys where the swift rivers of the Sierra Nevada flow, it is not unknown for a motorist to pull off the road by a mountain stream, snap off a wheel cover from his car, and wash out a few pansful of gravel.

The automobile wheel cover is probably about as efficient as a gold pan as the type used by the 49ers. Their's were often made of wood, carved from a wooden block by the prospector himself. Then came the hand-forged iron pan. Also the iron skillet. Many a skillet, says one historian, went straight from the breakfast campfire to the river. The next step forward was the steel pan. All of these had much the same size and shape. They were 12 to 18 inches in diameter, measured across the top, with flat bottoms and sloping sides.

Today, plastic pans are preferred. They cost $2 to $3, seldom more. They are light in weight, will never rust, and because they're black it's easier to spot a flake of gold. With a plastic pan, you can use a magnet to separate the black magnetic sands, called concentrate, from the gold particles you may have. When you draw the magnet over the outer surface of the pan bottom, it attracts the black sand, leaving the gold specks free.

Another advantage with plastic is that the pan's inner wall can be fitted with small ridges—called riffles—that run parallel to the rim and serve to trap gold particles as you agitate the pan material. This speeds up the panning process.

Beginners in the art usually use a pan that is 12 to 14 inches in diameter. Larger pans—16 to 18 inches—are for the experienced. They require a greater degree of skill and also more strength and stamina.

Panning is necessary to every type of small-scale mining. When you empty a sluice box, panning is still necessary to refine the black sand trapped by the riffles.

Modern gold pans feature riffles. *(Keene Engineering)*

A novice can learn the basics of panning in about a minute and a half, but doing it quickly and efficiently is an art that only the experienced acquire. When you pan, you are taking advantage of the fact that gold is seven or eight times heavier than the gravelly material in which it is found, and nineteen times heavier than water. The idea is to get rid of the lighter materials, but not lose any of the heavy black sand that possibly contains gold particles or, of course, any of the flakes themselves.

First fill the pan to one-half to two-thirds of its capacity with gravel and dirt. Squat at the water's edge and put the pan under water. Move your fingers through the dirt to be certain it all gets wet. Some experts suggest kneading the dirt just the way you would bread dough. The point is that you are going to lose any bits of gold that stay dry.

With the pan submerged, tilt it back and forth a few times. This is a preliminary step meant to size the material and also float off anything organic. In addition, it helps to start the gold flakes—if you have any—on their way to the bottom. Lift the pan out of the water and pick out any big pieces of gravel and toss them away.

Now you start getting serious. Begin swirling the water around in the pan, every now and then tipping the pan away from you so that some of the water and lighter rock material slops over the pan lip. Keep doing this, submerging the pan whenever you need more water. Keep swirling, letting water and coarser, lighter material slop over the side. Little by little (the whole process will take about 15 minutes), the material in the pan becomes finer and finer. You end up with a thin layer of black iron sand, maybe a couple of tablespoons of it.

What some people do at this point is add enough water to spread the sand evenly over the pan bottom. Then, using tweezers, they pick out any visible gold particles. These are put into a glass vial with a screw top.

Others prefer to probe the black sand concentrate at home. In the field, they pick out only the larger flakes, those $1/8$ of an inch in diameter and bigger. The black sand is then dumped into a container. At home, it's given a very careful examination. A teaspoonful at a time is placed in the pan, swirled about, then examined. This method enables you to find even the tiniest of flakes: microparticles. "I shudder to think how much gold I lost before I started handling my concentrates at home," says one veteran prospector.

## THE SLUICE BOX

Panning is hard work. That's why so many latter-day gold seekers rely on the sluice box.

A sluice box is an inclined trough which is placed in a fast flowing stream or river at a shallow depth. The water flows through at a fairly

First step in panning is to submerge the pan and knead the material to thoroughly wet it. *(State of Montana, Department of Highways)*

Operating a sluice box on a northern California stream. *(Carl Fischer)*

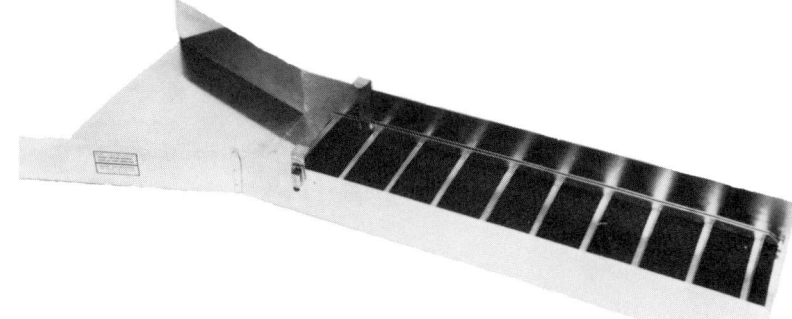

The Mini Sluice.
*(Keene Engineering)*

rapid rate. The bottom of the box has several transverse obstacles called riffles. When small amounts of gravel are shoveled into a hopper at the upstream end of the box, the material washes over the riffles, which serve to trap black sand and gold flakes. The lighter, unwanted material flows out the other end. Beneath the riffles, there is a layer of carpeting which captures even the tiniest of gold particles. In other words, the river does the work.

Sluice boxes can be of almost any length. In days gone by, miners used Long Toms, sluice boxes that were as long as a quarter of a mile, and required fifteen or twenty men for shoveling sand and gravel. Today's sluices are usually only a few feet in length. And they're very portable.

Besides the sluice, you'll also need a shovel and bucket. Use a small shovel, such as an Army trenching shovel. Its blade can get beneath and between boulders with greater facility than an ordinary garden shovel. Dig out the material that is to go into the sluice box and fill the bucket.

To be successful with a sluice box, you have to be very selective about what you put into it. Avoid sandy areas. Instead, look for spots that are heavily packed with gravel. Look in rock-strewn areas and underneath mossy banks. Look amidst tree roots and check out crevices in stream bedrock. Bring along a trowel to remove gravel and sand from these natural "gold traps."

When setting a sluice box in a stream, try to get about an inch of incline for every foot of the box's length. A 3-foot sluice box should have a 3-inch tilt. One way to know whether the water is flowing at the right velocity is to drop a small shovelful of material into the hopper and watch what happens. It should flow right out, not too quickly, and break up gradually, taking about thirty seconds.

From time to time as the water flows, pick out the larger pieces of trapped gravel and throw them away. Examine them carefully first! Garnets and other gemstones (see below) are often found in stream beds and they are well worth saving.

When the sluice box is filled to capacity with black sand, set it on a stream bank and remove the riffles. Scoop out the black sand with the trowel and put in a container. Then remove the carpeting and carefully wash what it holds into the container. You'll probably have to do this once or twice a day, and more frequently if you're in an area where there is black sand in large quantities. Later, perhaps when you get home, pan the black sand that you've recovered. You can do it while you're watching television.

After you're through panning the sand, don't throw it away. To commercial processors it's worth $1-$10 a pound. They're able to recover yet more gold from it with their sophisticated processing methods.

Sluice boxes are not expensive. For example, the Mini Sluice is a unit 36 inches long and 10 inches wide, the riffles are removable. Made of aluminum, it weighs 5 pounds and costs about $20. One that's bigger—52 inches by 10 inches—to give increased water velocity and greater capacity, costs $35. This model, also made of aluminum, weighs 10 pounds.

The benefit to be derived from using a sluice box is easy to understand. "A professional panner can process anywhere from one-half to one cubic yard of material a day," says Jerry Keene, head of Keene Engineering, a person who has been involved in small-scale mining for the better part of his adult life. "With a Mini Sluice, the average inexperienced person can process nearly ten times that amount."

## SURFACE DREDGES

At the next level you encounter the surface dredge, sort of a sluice box with a vacuum hose attached. You suck up sand and water from the crevices in bedrock or from the bottom of a river bed, and the material flows through the hose to a hopper and into the sluice. The running water carries away the light, unwanted material and later you pan what gets trapped. A dredge can handle a cubic yard of material in an hour and advanced models can handle several times that amount.

Surface dredges are based on a simple operating principle: Water under high pressure is pumped through a metal tube and introduced to the hose at a point just beyond the hose mouth. This creates a vacuum, a powerful suction, at the mouth. A gasoline engine like that of a lawn mower provides the operating power. Often the entire unit is mounted on a rubber-tube float.

It is important to use a dredge of appropriate size for the type of work you will be doing. Dredges are classified: 1½ inches, 2½ inches, 3 inches, etc. This rating refers to the inside diameter of the suction hose not, as many people believe, the diameter of the suction nozzle. The nozzle is from ½-inch to a full inch smaller in diameter than the hose (to help prevent the hose from becoming clogged.)

The diameter of the hose bears a direct relationship to the type of work the unit is meant to perform. A dredge with a 1½-inch hose is what you want for working crevices in stream bedrock. It's not for attacking a big sandbar or gravel bank. "The user of a small dredge," says Jerry Keene, "should confine himself to areas where the overburden gravel on top of bedrock is no more than four or five inches in depth."

The next largest size is the 2½-inch dredge, a big step forward in terms of what can be accomplished. A 2½-inch dredge will process four times as much material as the 1½-incher, which translates into about four cubic yards of material in an hour. "It doesn't take long for it to make a helluva hole," says one dealer.

Dredges are rated up to 6 inches in size, but, as a novice, it's not likely that you'll require anything larger than the 1½-inch or 2½-inch unit. Keep in mind that as the hose size increases, so does the weight of the dredge and its fuel requirements. The 1½-inch dredge weighs about 25 pounds and some require no more than a gallon of gasoline for every six hours of operation. With a 2½-inch unit, the weight is about 70 pounds and fuel consumption is a gallon of gasoline every three hours. Weight and fuel consumption more than double when you talk in terms of dredges rated at 3 and 4 inches.

Dredges of 1½ and 2½ inches are eminently portable, enabling you to backpack the unit to remote mountain streams that haven't been worked over in years. They're not expensive, either. Keene Engineering offers a 1½-inch unit (Model 1502) equipped with 7 feet of suction hose, 6 feet of pressure hose, and a 1 hp Olsen and Rice 2-cycle engine. The sluice box and flotation platform are plastic, molded into a single unit. The dredge weighs 28 pounds; it costs $195.

If you can run a power mower, you can operate a surface dredge of this type. You start the engine by pressing a primer button on the carburetor and yanking a starter rope.

Keene's 2½-inch unit (Model 2501) is powered by a 3 hp Briggs and Stratton engine. It comes equipped with 10 feet of 2½-inch vinyl suction hose and 12½ feet of pressure hose. The entire unit is mounted on a metal frame that is fixed atop a heavy duty inner tube. It weighs 77 pounds; it costs $319.50.

Surface dredges are sometimes used in conjunction with diving gear. With the simplest diving equipment—a snorkel and a facemask—you can operate in water that's three or four feet in depth. The diver glides downstream slowly, the dredge in tow, cleaning out crevices in the bedrock with the nozzle. Material deep in crevices is reached with a small pick. Of course, for deepwater diving, a scuba rig is necessary.

## DRY GOLD

While the wet-washing methods of gold recovery are the best known and more popular, a growing number of prospectors prefer working desert areas and using dry techniques. When using sluices, dredges, and such, they say, you're restricted to a limited number of rivers and streams, many of which have already been thoroughly worked over. To get to areas that haven't been covered, you must trek deep into the back

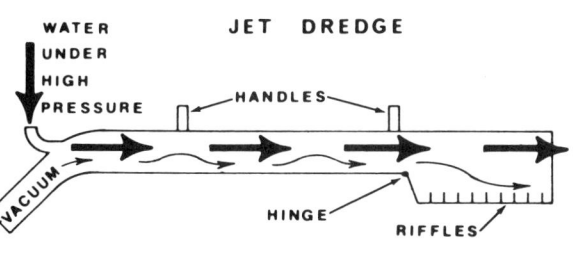
*(California Division of Mines and Geology)*

Surface dredges are portable, easy to operate. *(Keene Engineering)*

country, which is no picnic when you're packing along a dredge and all the related equipment.

Desert areas, dry gold prospectors and miners point out, are more accessible. "It's often possible to drive your camper within a few feet of the spot where you're going to operate," says Carl Fischer, a veteran of this type of mining. "Sleeping in a good bed in a warm camper sure beats a confining sleeping bag on the cold ground."

In California, the most productive areas for dry placer mining are to be found in the El Paso Mountains, Chocolate Mountains, and the Picacho region of Imperial County, which borders Mexico. Mexican and Indian laborers worked these areas more than a century ago. Many of the arid regions of Nevada, Arizona, and New Mexico are also potentially productive.

Where streams once existed, or now run intermittently, is the best place to look. Once you've found a dry wash that seems promising, go over a good-sized chunk of it with a metal detector, a BFO unit, looking for black sand concentrations. Don't stop to dig at the first one. Mark it and go on to look for others. Dig out the sand that gives the strongest reading. When you reach bedrock, don't fail to clean out the crevices and potholes.

For experienced prospectors and miners, there's this advanced model. *(Keene Engineering)*

The Model 1502. *(Keene Engineering)*

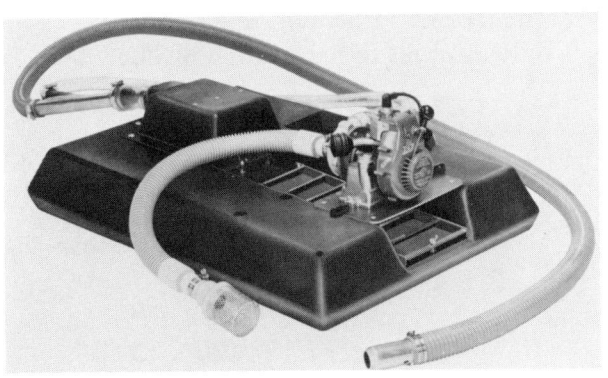

There are two basic methods of recovering dry gold from sand and gravel: with the dry washer or the electrostatic concentrator. Both units are available from mining equipment companies.

The bellows type of dry washer is very popular. It works like this (see illustration): Sand and gravel are shoveled onto a screen at the top of the washer (A), and the finer material drops through into a hopper (B). From the lower end of the hopper, the material falls into a set of riffles (C). Air from a bellows (D) lifts the lighter particles of material off the lower end of the washer, while the heavier particles—black sand and gold flakes—remain trapped behind the riffles.

The dry washer can be operated by a hand crank or gasoline engine. Such units cost $100 to $200, depending on whether they are motorized.

With some dry washers, you can control the flow of material from the hopper to the riffled tray. "We make it a rule to stop the machine every hour to check the deposits in the tray," says Carl Fischer. "We've found that most of the gold is deposited in front of the first and second riffles."

In days past, dry gold was sometimes obtained by a process known as winnowing. Gold-bearing sand and gravel would be tossed into the air from a blanket. The heavier particles would fall back, while the wind carried away the unwanted lighter materials. Often a wool blanket was used because an electrostatic charge would build up in it, which helped the blanket to retain the gold particles.

The electrostatic concentrator is based, at least in part, on this principle. The concentrator resembles a dry washer in size and general appearance and consists of a pair of tilted trays, one mounted above the other so as to feed into it. An air fan, powered by a gasoline engine, is used to create an electrostatic charge within the bottom tray. This tray is equipped with riffles, and beneath the riffles is strung a cloth of synthetic material. This receives the electrostatic charge.

Gold-bearing sand gravel is shoveled into the top tray, which is equipped with a trommel screen to sift out the larger material, allowing only the smaller particles to pass into the concentrator tray on the bottom. As the material moves through the riffles, the gold particles, even though nonmagnetic, have an affinity for the charged cloth and adhere to it. As much as two tons of gravel can be processed with such a unit in the space of an hour.

The electrostatic concentrator can be folded onto a portable package weighing about 80 pounds. It takes only minutes to reassemble it. The unit pictured here costs $295.00.

Operating a dry washer.
*(Carl Fischer)*

Detectors serve in dry-gold prospecting, too. *(Carl Fischer)*

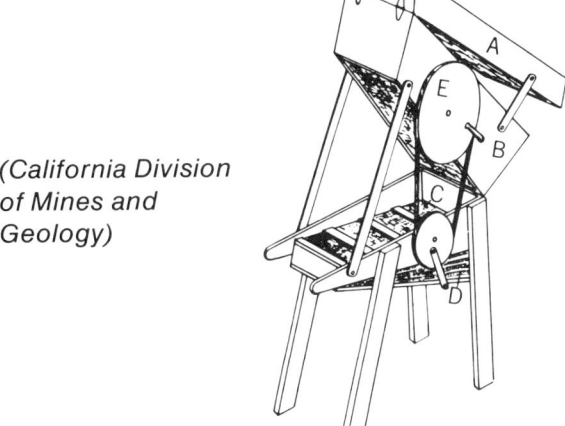

*(California Division of Mines and Geology)*

Check the tray carefully.
*(Carl Fischer)*

## ABANDONED MINES

In the states west of the Mississippi, there are, according to some official estimates, approximately 20,000 abandoned mines. The reason they've been forsaken is that it is no longer economically feasible to work them. Yet these mines, and the piles of tailing usually to be found all about, are often well worth investigating.

Some mining companies were after a specific kind of mineral or metal and the waste they left behind may contain many worthwhile minerals. Also bear in mind that mining methods of the past were often inefficient; there were no electronic devices to help spot finds.

A word of caution first. Old mines can present hazards. The shafts are sometimes concealed by thickly grown brush and vines. The tunnel timbers may be decayed and weak. The perils are obvious.

There are also legal dangers. Often mine property and the mines themselves are not fenced or posted to indicate ownership. Yet trespassing on private land, whether it is posted or not, is in violation of the laws of most states. "Many owners of land, especially owners of mining property, do not want others removing gem stones or mineral specimens," says a spokesman for the Denver office of the Bureau of Land Management. "Serious disputes sometimes result when a person wanders onto private land and starts digging without permission." The solution is to obtain permission ahead of time.

Searching old mines requires special techniques. The tunnel floor is the best place to look. When the mine was in operation, all the high-grade ore had to be hauled out of the tunnel to be milled. Whether it was hauled by hand or ore cars, some ore was dropped. These high-grade ore samples, now covered by mine debris, are what to look for.

Use a BFO detector with a 3½- to 5-inch loop when searching for these ore samples. But the methods of search are not standard ones. Lay the detector on the ground. Tune it in so that you're getting a slower-than-normal beat. Use a rock hammer and dig down into the tunnel floor beyond the layers of waste material. When you find an ore sample that seems promising, pass it under the search loop.

Remember, the samples that you find aren't going to be large ones. Anything big would have been spotted by the miners. But because prices have mushroomed so, the specimens don't have to be sizeable ones. "Mine-floor searching can be a very profitable venture," says Charles Garrett, "and the chance of coming home empty-handed is small indeed."

When working mine tailing or a mine dump, use the same method of search as described above. Set the detector flat on the ground and after

Abandoned buildings at the Union Companion Mine, Cornucopia, Oregon. *(Oregon State Highway Department)*

Examining a mineral specimen near Bluff, Utah. *(Utah Travel Council)*

recovering promising samples, pass them under the search loop. There's simply too much mineralization about to work in normal fashion. The best method is to begin digging in random fashion, taking samples from several different sites until you find one that seems promising.

STAKING A CLAIM

If you find an area rich in mineral deposits, you may want to take the next step: staking a claim to the site. A claim gives you the exclusive right of developing and extracting discovered minerals.

Gathering specimens at the Sheep Creek Agate Beds near Prineville, Oregon. *(Oregon State Highway Department)*

How you go about filing depends on who owns the land in which you're interested. If it happens to be private land, then it's a matter of negotiating with the landowner, or the holder of the mineral rights to the property. If the land is state-owned, you proceed in accordance with state law. Write to the State Land Commission for information.

In the case of national forest land, or the 450 million acres under the jurisdiction of the Bureau of Land Management, certain specific procedures have been established for filing a claim. In such cases, you also must comply with the regulations set down by the state involved. Most of the public-land states, particularly those of the West, publish pamphlets containing their regulations.

The first element in establishing a claim is the actual discovery of a valuable mineral. (The term mineral refers to both metallic and nonmetallic substances. A mineral can be either gold, silver, lead, etc., but it can also be asbestos, mica, fluorspar, etc.) In other words, you can't file a claim for a site with the expectation that you're going to find something worthwhile. The federal government states that claim sites must be an area "where minerals have been found and the evidence of such character that a person of ordinary prudence would be justified in further expenditure of his labor and means, with a reasonable prospect of success, in developing a valuable mine...."

Once the find is made, the next step is to mark the boundaries. There is some variation from state to state as to how this is to be done, but it usually means erecting posts or stone monuments at the four corners of the claim site. In the case of placer claims, which are located by legal subdivision, corner markers are usually not required. Most states mandate that you post a notice of location at the claim site, usually at the place of discovery.

Then you must record the claim. At the county recorder's office in the county where the claim site is located, you place on file the following information: name of the claim, type, mineral claimed, the name of the locator, and a description of the site.

Generally, the procedures outlined above also apply to national forest wilderness areas. In such cases, you must also file a written notice of the location of the claim with the appropriate district ranger or forest supervisor.

Once you're in possession of a claim, you must work it in order to keep it. The Federal government requires that you perform at least $100 worth of labor, or improve the claim to that extent, each year. There are state requirements in this regard, too. The annual labor is known as assessment work. Statements as to what it consists of and when it was performed must be filed with the county recorder annually.

There is no limit to the number of claims an individual may hold, as long as each concerns the discovery of a valuable mineral and the other requirements are fulfilled. All of the work involved in marking the boundaries, posting the notice of location, and recording the claim, should be carried out with meticulous care. The more diligent you are, the less likely that your claim will be appropriated by someone—be "jumped."

You may establish a valid claim, perform the annual assessment requirements and, thus, hold certain possessory rights, but you still do not hold full title to the land. To achieve ownership, you must secure a patent.

An application for a patent is made through the Bureau of Land Management. Since the procedure is rather a lengthy one involving numerous legal documents, most applicants use attorneys.

Applying for a patent involves an official survey of the claim (usually one of the chief costs involved), the filing of papers showing proof of possessory right, the payment of the filing fee, the publication and posting of filing notices and, ultimately, the payment of the purchase price. The price is $5 an acre for quartz claims and $2.50 an acre for placer claims. The cost of the land is the smallest incurred during the proceedings.

For more information on securing patents, write an office of the Bureau of Land Management (see addresses above). Keep in mind that a patent is not necessary in order for you to continue to hold the rights established by your claim, and that those rights can be leased, sold, or willed to heirs.

Additional information on claim filing can be obtained from the regional offices of the Bureau of Land Management or Forest Service (see above), or from Forest Supervisors. These are among the booklets available: *Staking a Mining Claim on Federal Lands, Patenting a Mining Claim on Federal Lands, Mineral and Mining Claims in the National Forest Wilderness,* and *Regulations Pertaining to Mining Claims.*

## FINANCIAL ASSISTANCE

The federal government, in an effort to stimulate exploration for gold and other minerals, has established a program of financial assistance for qualified individuals. Because you have to own or lease the property you plan to explore and provide 25 percent of the funds yourself, it can't be considered an out-and-out grubstake, but there are some very tangible benefits. The principal one is that the government will loan as much as 75 percent of your costs of exploration, and if you don't happen to find any gold or whatever else you happen to be looking for you don't have to pay back the money.

To qualify for the program, you have to furnish evidence that funds for the exploration are not available to you from any bank or other commercial source at reasonable terms. You also have to show that you would not ordinarily undertake the proposed exploration at your own expense.

The government's share of the costs is paid monthly after a voucher for costs incurred the preceding month has been submitted and approved. Persons receiving the loans agree to pay the government a 5 percent royalty on whatever proceeds are derived from the property. "There is no requirement to produce," states a government bulletin on the subject, "and if there is no production no repayment is required."

Approximately $5 million has been loaned out under the terms of the program, and the government has been repaid about 10 percent of that figure. "However," says a government spokesman, "for each dollar the government invests, the value of reserve minerals found is estimated at $30."

Gold isn't the only mineral covered by the program. The government will also loan up to 75 percent of your exploration costs if you're searching for mercury, silver, tin, and platinum-group metals, and 50 percent of the costs when exploring for asbestos, bauxite, beryllium, cobalt, manganese, mica, nickel, thorium, and uranium. This is only a partial list.

For more information, write to the Field Office of Minerals Exploration, U.S. Geological Survey, 345 Middlefield Rd., Menlo Park, CA 94025. Ask for the booklet titled *Exploration Assistance*—it's free—and an application form.

## FURTHER INFORMATION

Countless informative and authoritative pamphlets are available on the topics of gold prospecting and small-scale mining. Many are free. Here is a listing of some worth writing for:

*Suggestions for Prospecting.* A well-written introduction, includes a comprehensive bibliography. Order from the U.S. Department of Interior, Geological Survey, Washington, DC 20242.
*Production Potential of Known Gold Deposits in the United States.* Publications Distribution Section, U.S. Bureau of Mines, 4800 Forbes Ave., Pittsburgh, PA 15213.
Bureau of Mines Information Circular 7695, "Laboratories That Make Fire Assays, Analyses, and Tests of Ores, Minerals, Metals, and Other Inorganic

Substances." A listing of private concerns in the United States and Canada where mineral samples may be sent for identification and other testing to determine their value. Publications Distribution Section, U.S. Bureau of Mines (see address above).

*Placer Evaluation and Dredge Selection.* Superintendent of Documents, Government Printing Office, Washington, DC 20402; cost, 60 cents.

Virtually every state with gold deposits has available information for fledgling prospectors and miners. It is frequently extremely detailed and authoritative, having been prepared by state geologists. The material offered by California is especially noteworthy; it can't help being valuable. Consult the appendix of this book for information as to what state department or division to write to, and what to ask for.

You can also obtain helpful information in the form of catalogs from the companies that sell equipment for prospecting and small-scale mining. Keene Engineering (11483 Vanowen St., North Hollywood, CA 91605) is the largest firm of this type. Miners (Riggins, ID 83549) is the name of one of the foremost suppliers in the field of geological and prospecting gear.

## FINDING GEM MINERALS

Admittedly, your chances of finding gold in really worthwhile quantities are going to be limited unless you happen to live in the western states or plan to visit there. But if you happen to live east of the Mississippi River, don't lament. There are plenty of opportunities for prospecting. Instead of gold, you hunt for gemstones.

A gemstone is a precious or semiprecious stone that may be used as a jewel when cut and polished. In other words, a gemstone is what precedes a gem.

Every state has gemstones of several different types. The underlying rock of a region frequently determines the type of gem materials that will be found. Climate is believed to be a factor, too. Turquoise is found in parts of the world where rainfall is limited and, thus, is common to the American southwest. Topaz of gem quality has been found in California, Colorado, Texas, Maine, and New Hampshire. Tourmaline in varying degrees of quality has been found in just about every state, but the most valuable occurrences exist in two widely separated areas: California and the Northeast. Quartz and opal, both silica gemstones, are to be found in many, many regions of the United States. The same is true of petrified wood. A listing of localities in the United States where gem materials of different types are to be found appear in the appendix.

Of the approximately 1,500 mineral species, only about 100 can be classified as gems. Of these, only 4 are regarded as precious: diamonds, rubies, sapphires, and emeralds.

*Diamonds* The superstar of gems, the diamond is the hardest of all substances. By virtue of its brilliance and clarity, it has come to be prized in every part of the world. The first diamonds came from India, and most noted stones in history are Indian in origin. As India's production of diamonds began to dwindle, diamonds were discovered in Brazil.

Then, in 1867, one Schalk van Niekirk offered to buy an unusual stone he saw in a farmhouse in South Africa. It was only a child's plaything he was told, and it was given to him. What happened is well known to history. A mineralogist identified the stone as a diamond worth several thousand dollars, and thus began the great diamond rush. Today, South Africa supplies about 90 percent of the world's diamonds, both those used in jewelry and for industrial purposes.

Are diamonds to be found in the United States? Indeed. They have been recovered in the Great Lakes states, particularly Wisconsin, Michigan, Indiana, and Ohio, although no discoveries have been made in recent years. Geologists theorize that as the mighty glaciers ground south across what is now Canada, they displaced diamond-bearing rock formations, volcanic in nature, that were located somewhere near Hudson Bay. Other of these diamonds are believed to have been found in Kentucky and Tennessee.

Random diamonds have also been recovered amidst the sand and gravel of present and former stream beds in North and South Carolina, Virginia, Georgia, Alabama, West Virginia, Texas, Idaho, Montana, Oregon, and California. Discoveries in the West have usually been made in connection with placer-gold mining efforts. It is believed that the original source of these diamonds were the once volcanic regions of the Rocky Mountain range. Streams and rivers carried the diamonds downstream to the sites where they were found. In the East, it was the Appalachian range that was the original source.

The one area in the United States where diamonds have been found in what might be called their natural state is near Murfreesboro, Arkansas. Here a massive underground volcano erupted, forcing lava up through layers of shale. The carbon present in the shale was crystallized by the heat and pressure into diamonds. More than 55,000 diamonds have been taken out of the clay and dark soil that forms this now famous open mine.

This area, which covers 74 acres, is open to amateur prospectors. Upon payment of a nominal fee—$2 for adults, $1 for children—you're permitted to hunt for a full day. Some families spend their vacations there. Murfreesboro and nearby Lake Greeson boast fine motels and large and modern campgrounds. The owners of the site deeply plow the area from time to time, so that there's fresh earth to probe. Some stones have been found that have been 20 carats and more in weight.

Unless you're searching the Arkansas mine and have a clear idea of precisely what you're looking for, you probably wouldn't recognize a diamond in the field. The best thing is to visit a local museum where rough diamonds, those not cut or polished, are on display.

A rough diamond has no gleam to it. If found in a stream bed, it often looks like a piece of quartz coated with a grayish film. A lapidary appraising the stone cuts a minute "window" in the coating to get a look at what's inside.

As for size, most diamonds are so small that it is not likely that they will ever be recovered, although diamonds found in stream beds in the United States have been as large as 5 carats in size. The Uncle Sam diamond, the biggest ever found in this country, weighed 40.23 carats.

Prospecting for diamonds is never going to rival bike riding as a leisure time activity, but there are thousands of people who enjoy the hobby. When working stream gravel for diamonds, the principal piece of equipment you need is a sieve, or a series of them. Gordon S. Fay, in his book, *The Rockhound's Manual,* suggests using a screen of ¼-inch mesh first, and then further sifting the material with ⅛-inch mesh. Don't be in too much of a hurry to throw away the "worthless" material from the "reject" screen, he says, and don't discard the material that passes through the finer screen. Both batches of material should be examined for gold particles.

*Rubies and Sapphires*   Carat for carat, a good-sized ruby is worth more than a diamond of the same size. A 5-carat ruby is twice as valuable as a 5-carat diamond, and a 10-carat ruby has three times the value of a 10-carat diamond. The reason is that large rubies are extremely rare. Small ones are not quite so scarce and, thus, have proportionately less value.

Ruby and sapphire are both varieties of the same species of mineral: corundum. The chief characteristic of corundum is its hardness; it is second only to diamond in this regard. When corundum is red in color, it is known as ruby. If it's any other color, it's sapphire. Most people think of sapphire as blue, but it can also be green, purple, pink, and even golden. Then there's the star sapphire, extremely popular in recent years. When properly cut, it presents a six-rayed starlike optical figure.

Rubies have been mined commercially in North Carolina, and sapphires in the Yogo Gulch area of Montana. There have been random findings of either or both of these stones in about fifteen different states.

Ruby and sapphire are formed in what are known as metamorphic rocks, such as marble, schist, and gneiss, and also in some igneous rock, particularly those having a relatively low silica content. Rather than attempt to find the stones in the host rock, the usual method is to pan for them in the streams and rivers below the rock formations. As in the case of diamonds, fine mesh screens are frequently used.

*Emeralds* In mineral and gemstone handbooks, the emerald is classified as one of several gems belonging to a mineral species known as beryl. Other varieties of beryl are aquamarine, morganite, and goshenite.

Emerald is softer than the others, but color is the feature that rockhounds use to distinguish one from the other. Emerald, as is well known, is green, grass green. Aquamarine is, well, aquamarine; it ranges from green blue to blue green. Morganite goes from pink to rose red. Goshenite, named after Goshen, Massachusetts, where it was first found, is colorless, resembling milky quartz.

What Murfreesboro, Arkansas, is to domestic diamonds, Hiddenite, in the western part of North Carolina, about 25 miles west of Winston-Salem, is to emeralds. In 1879, one William Earl Hidden, a geologist in the employ of Thomas Edison, was exploring the region in search of platinum when he came on a number of beryl-crystal gemstones, not emeralds, but a variety of gemstone not previously known. The mineral was named hiddenite in his honor as was the village that sprang up near the site of his finds. Later, Hidden did find emeralds, but the really important discoveries there were of a recent nature.

In 1969, an amateur rockhound named Michael Finger found a dazzling emerald not far from Hiddenite. It was 2 inches in diameter and 3 inches long, the largest emerald ever found in the United States. When cut, it brought more than $100,000.

Emeralds are now mined commercially in this region, and there are also abundant opportunities for amateurs. Two mines permit gem hunters to do unlimited digging, sifting, and panning at the rate of $3 a day.

One clue to emeralds in this area are mica veins. Below the mica there is usually quartz, and emeralds and other gemstones sometimes crystallize beneath the quartz. Dirt underneath the quartz is shoveled out and sifted.

Emeralds have also been found in South Carolina and Massachusetts. It is not unlikely that they occur in several other areas, as yet undiscovered.

## NECESSARY EQUIPMENT

No matter what types of gems and minerals you're after, the work you do in their pursuit will fall into three basic categories: breaking, sorting, and transporting. The tools you need to accomplish these chores are as follows:

*Rock Pick*  This is almost as basic to rockhounding as skis are to a skier. The word pick doesn't quite describe it; the tool is actually a combination hammer and chisel, expertly designed and carefully balanced. Choose one that has a comfortable feel to it. The size of the handle and the weight of the head—from 14 to 22 ounces—are the features to evaluate.

A premium rock pick, with a 22-ounce, polished head and shank, and a laminated leather grip costs between $7 and $8. To carry the pick, buy a leather belt-sheath or holster. One costs only a few dollars.

*Chisels and Gads*  Chisels have beveled ends and are used to split rock along a specific directional line. Gads, which have pointed ends, are used to force rock apart. Chisels and gads cost $2 to $3 apiece. Look for the type with vinyl grips. Chisel holsters are available, too.

*Hammers*  The rockhounding hammer is the sledge type with polished faces. Heads can be 3, 4, or 7 pounds in weight. The heavier the hammer the easier it is to break apart rock samples, but keep in mind that the careless use of hammers has ruined countless worthwhile samples. Hammers cost $8 to $9.

*Sample Bags*  Dealers in prospecting and geological supplies also sell cloth bags to transport your rock samples from the field. There are both cotton cambric and canvas bags, the latter for the heavier samples. The cambric bags are $6.70 for a hundred. Canvas bags, $25.75 per hundred, can also be purchased in lots of 25.

A quality metal/mineral detector can be an important accessory in your search for gem materials. A BFO machine is superior to a TR unit in this regard, because the latter may send off signals that will confuse you. No matter which type you use, test the unit and its search coil to familiarize yourself with the responses it will produce. Use rock specimens of the type you expect to be finding, passing each under the search coil several times.

## ADDITIONAL MINERAL INFORMATION

You can do many things to become more informed about the minerals in your area. You can join a rockhound club. There are about a thousand of these in the United States.

Most local clubs are affiliated with the American Federation of Mineralogical Societies (3418 Flannery Lane, Baltimore, Maryland 21207), whose six regional associations boast over 50,000 members. The aim of the organization is to promote popular interest and education in the various earth sciences, particularly geology, mineralogy, paleontology, lapidary, and related subjects.

Every community has a rock shop where information is available. These are listed in your Yellow Pages under "Minerals." Besides selling equipment for rockhounding and individual rock samples, these shops often sell complete mineral collections containing specimens gathered by local rockhounds. Each specimen is about the size of a matchbox. And each is numbered and keyed to an identification sheet. These can be a big help when it comes to field identification of the samples you'll be finding.

The geology departments of local colleges can be helpful, too. Professors and students are sure to have a thorough knowledge of the geological makeup of the area. Local museums are likely to have on display minerals that are common to the region.

Excellent books are available on the subject of gemstones. They include the following.

Fay, Gordon S. *The Rockhound's Manual.* New York: Harper & Row, 1972, $7.95. A clear, concise, and authoritative introduction to rockhounding. The author is a geological engineer and a one-time U.S. mineral surveyor.

Kunz, George Frederick. *Gems and Precious Stones of North America.* New York: Dover Publications, 1968, $4.00. First published in 1892, this is regarded as the

Test rock samples to determine the type of signals you're going to get. *(George Sullivan)*

definitive book on the subject. Its list of gemstone locations is perhaps the most comprehensive available. Highly recommended.

MacFall, Russell P. *Gem Hunter's Guide.* New York: Thomas Y. Crowell Co., 1969, $6.95. A state-by-state listing of gemstone locations.

Sinkankas, John. *Gemstones and Minerals: How and Where to Find Them.* New York: Van Nostrand Reinhold Company, 1961, $17.50. Informative chapters on mineral identification and various aspects of prospecting: maps, tools and other equipment, and fieldwork.

A number of fine magazines are devoted to rockhounding:

*Earth Science* (P.O. Box 550-GM, Downers Grove, IL 60515). The official magazine of the Midwest Federation of Mineralogical and Geological Societies. One year, six issues, $3; sample copy; 50 cents.

*Gems and Minerals* (P.O. Box 687, Mentone, CA 92359). The official magazine of the American Federation of Mineralogical Societies, the California Federation of Mineralogical Societies, The Northwest Federation of Mineralogical Societies, and the Texas Federation of Mineralogical Societies. One year, 12 issues, $3; sample copy; 35 cents.

*Rocks and Minerals* (Box 29, Peekskill, NY 10566). The official magazine of the Eastern Federation of Mineralogical and Lapidary Societies. One year, 12 issues, $3; sample copy, 60 cents.

Don't overlook the Bureau of Mines and Mineral Resources, or the Department of Geology, for the state in which you plan to search. Most states publish extremely informative booklets for rockhounds. Consult the listing of state information sources given in the appendix.

Chapter 7

# UNDERWATER SEARCH

The modern-day treasure hunter frequently wears fins and a facemask, and the waters offshore the length of Florida's east coast are likely to be where he does his prospecting. It's true, of course, that the bulk of the riches recovered from this region in recent years has been found by professionals, but it's also true that your chances of finding gold here are better than they were for the Klondike diggers of the 1890s.

This applies not just to Florida's coastal waters, but also to the Great Lakes, the Texas Gulf, along the much traveled shipping lanes between Boston and Philadelphia, offshore North Carolina—just about anywhere. The odds are more in your favor than they have ever been because of recent developments in underwater detection equipment, "recent" meaning within the last ten years. Hand-held detectors, which cost less than you might pay for a scuba outfit, have opened new horizons for anyone seeking random coins and relics buried in the sand near wreck sites.

This chapter presupposes that you are trained and conditioned for scuba diving, that you own the proper equipment, and have been schooled in how to use it. If this isn't the case, then you should make arrangements to take a course of instruction in scuba, one offered by an instructor who has won proper certification. Don't expect to learn how to dive by reading a book or by a trial and error process.

Several organizations conduct instructor certification programs: the National Association of Underwater Instructors (NAUI), the oldest and largest organization of this type; the National Association of Skin Diving Schools (NASDS): the National Council of YMCAs, and the Professional Association of Diving Instructors (PADI). Those experienced in the field say that the instruction programs offered by NAUI and the NASDS are likely to be more thorough than the others.

For information about instruction in your area, simply contact a local dive shop. These are listed in the Yellow Pages under "Divers'

Equipment and Supplies." Often dive shop listings in the telephone directory tell whether instruction is offered and, if so, the type of certification the program has. You can also inquire at a local YMCA.

Usually these courses involve 10 to 12 instruction sessions, each lasting a minimum of 2 hours. Part of each class is devoted to classroom lectures, part to inpool training. Tuition is usually $50 to $75. The student is expected to furnish his own mask, fins, and snorkel, but usually the basic scuba gear—air cylinder, filler valve, and regulator—are available at no additional charge. Be sure to establish that the course includes, in addition to the lectures and pool instruction, at least one "open water checkout," as it's called, that is, that you'll be instructed in the use of the scuba gear in an authentic diving situation, a lake or the ocean.

A good resume of what you will learn from such courses can be obtained from the book, *The New Science of Skin and Scuba Diving*, which is the textbook for the YMCA course. It is available at most bookstores or can be obtained from the National Council of YMCAs (291 Broadway, New York, NY 10017). The manual costs $2.75.

### WHERE TO LOOK

According to the Florida Board of Archives and History, there are between 1,200 and 2,000 sunken shipwrecked vessels along the state's coastline. The Board has detailed information on 250 of them, all of which are believed to contain gold and silver in important quantities.

The reason that such an abundance of sunken treasure lies off the Florida coast has to do with the policies of the Spanish government more than 400 years ago. From 1503 to 1790, an agency called the Casa de la Contracion, with headquarters in Seville, controlled trade between Spain and the New World. To protect cargoes from pirate ships or those of enemy nations, the Casa established a convoy system. Ships carrying gold and silver bullion and coins assembled at Havana, departing in groups of 10 or more. They would head north and west through the Straits of Florida, ride the Gulf Stream along the Florida coast until they reached the Carolinas, whereupon they would pick up the westerlies to Europe.

Tropical storms today are detected by weather satellites as they begin to form, but in the eighteenth century there was no such early warning system. One hurricane struck the 1715 convoy, sending 10 of its 11 vessels to the bottom. Eight of these are known to have foundered on the reefs off of what is today the city of Fort Pierce. A similar tragedy befell the convoy of 1733. This time the hurricane struck further to the south. Of the 21 vessels, 17 were lost, most of them piling up on the reefs amidst the Florida Keys.

It was no secret that millions of dollars in gold and silver were scattered off Florida's coast. As early as 1920, bathers were finding random Spanish coins on the beaches north and south of the Fort Pierce inlet.

The first serious attempts to get at what was down there came in the 1950s with the development of scuba gear. No longer was a diver encumbered with bulky equipment and a long air hose. He could glide about effortlessly, the tanks strapped to his back providing his oxygen supply, the air regulator at his chest automatically compensating for the vagaries in pressure.

Nevertheless, probing the vessels off Fort Pierce and elsewhere remained a hit-and-miss proposition. A couple of divers would meet up with a fisherman in a waterfront bar and, in return for a fee or a share of what might be recovered, he would tell them the location of what he thought to be a wreck. If the divers were able to find the hulk, the principal piece of gear they had for working the site was an "airlift," a sort of underwater vacuum cleaner. But the airlift method was slow and tedious, for the device had no sense of discrimination. It simply brought *everything* to the surface.

The big breakthrough occurred during the 1960s. A group of amateur divers, known today as Real 8 Co., located most of the Spanish vessels that had been wrecked in 1715. One reason they were successful was because they used magnetometers, instruments which are capable of locating masses of ferrous metal by measuring the intensity and direction of the magnetic fields they generate. In three years of salvage work, the Real 8 divers recovered more than 66,000 gold and silver coins, and artifacts of all kinds—table silver, gold toothpicks, and pewter plates; cannons and cannon balls. Estimates as to the dollar value of what they recovered vary, but $3,500,000 is a figure that is frequently used.

The discovery of new wreck sites continues to this day. Not long ago, Treasure Salvors, Inc., an enterprise that has recently won rank as the world's most successful group of treasure seekers, announced that it had found the graveyard of a Spanish fleet that had foundered in 1622 with up to $600 million in gold, silver, and bronze artifacts. Years of salvaging are planned.

All of this does not mean that the coastal waters of Florida are the only place underwater treasure seekers have at their disposal. Not at all. Sunken ships are to be found all along the coastal shipping lanes of the United States. Countless wrecks lie in waiting on the coral reefs of the Caribbean. There's enormous potential in the St. Lawrence River and the Great Lakes. Some sources say that 12,000 vessels have been wrecked in the Great Lakes water alone; others say the number is 15,000.

As many as 2,000 sunken ships may lie in the waters off Florida's east coast. *(Florida State News Bureau)*

Revealing research has been conducted in the area recently by the Department of Natural Resources of the state of Michigan. The state is seeking to establish an underwater park system, one that would involve many of the shipwrecks in Michigan waters of the Great Lakes. As part of its planning, the state has begun to inventory sunken vessels. The project is being supervised by Dr. Richard Wright of Ohio's Bowling Green University. By the end of 1973, Dr. Wright had gathered information on 600 wrecks in Michigan waters, largely vessels which had gone down from 1820 to the Civil War. He estimated that there were probably as many as 6,000 wrecks solely in the waters under his scrutiny.

An estimated 900 vessels have gone to the bottom off of Cape Hatteras. The Graveyard of the Atlantic is what they call the region. There are 4 Spanish galleons there, 1 bearing an estimated 400,000 pieces of eight. It went aground off Ocracoke Island in 1750 and has never been discovered. There are 40 or so Confederate blockade runners in the area, 4 German U-boats, as well as Allied tankers sunk by German submarines.

The locations of hundreds of wrecks are well known to local fisherman and charter-boat skippers. Because diving facilities in the area are limited, and likewise divers, only a handful of wrecks have been explored.

**RESEARCHING A SITE**

Once you become interested in a particular region, what you do in the way of research is limited only by the time you have available and the amount of money you're willing to spend. For example, Treasure Salvors, Inc., were aided by a historian, Dr. Eugene Lyons, who spent three years in Spain studying Spanish naval records. When a 63-pound bar of silver was salvaged, Dr. Lyons, by means of several identification marks on the bar, was able to establish the name of the vessel that was carrying it, and the fact that it was one of nine ships that sailed from Havana with 27 tons of silver and several tons of gold.

You probably cannot afford to hire and maintain in Madrid a historian of your own, but you can do much to establish a wreck site that offers potential. If you were to walk down the main street of Fort Pierce, Florida, or the main street of almost any coastal city in the state, stopping people at random and querying them as to where treasure wrecks might be located, every third or fourth person would give you at least some information. Local fishermen know where wrecks are, often very precisely. If you inquire at local dive shops, you're sure to get any number of tales of salvage attempts of the past in the area, both failures and successes.

However, there's no reason you can't begin with a contemporary source: a newspaper account of a shipwreck, a magazine article, or a book. In the category of books, perhaps the best is *The Treasure Diver's Guide,* by John S. Potter (Doubleday & Co., 1960, $15). Close to 600 pages in length, it is the most comprehensive source available on the subject of where to search, that is, the locations of sunken vessels. A new edition was recently issued; before its publication, the original edition was reportedly selling for as much as $50 in secondhand bookstores. Other helpful books of this type are listed in the appendix. The matter of researching by means of magazines and newspapers is covered in Chapter 3.

Newspaper stories that appeared at the time of the vessel's sinking can be of help. Keep in mind that libraries in large cities often maintain microfilm or bound copies of major foreign newspapers. Sometimes these files date back a century or more.

Professional divers never fail to seek some official report of the wreck. In the case of a loss that is of a fairly recent nature, they will try to get the ship's manifest, its listing of cargo and passengers. Sometimes this can be found in the files of the shipping line, or the agent at the port where the cargo was put aboard may have a copy. Insurance company records are also consulted. These will describe the cargo in detail and place valuations on it. Since its founding in 1688, Lloyd's of London, either through its home office or one of its more than 300 brokers around the world, has insured many thousands of oceangoing vessels. Unfortunately for researchers, company records were destroyed by fire in 1838. What information that does exist begins with that date.

Professionals also seek out state papers from government archives, especially in the case wherein one government may have communicated with another concerning the ship's cargo. The naval archives of many countries often contain detailed descriptions of naval engagements. Court-martial records are still another source. In the British Navy it was standard procedure to bring to trial any captain who was deemed responsible for the loss of a ship.

Both professional and amateur treasure seekers look to the federal government as a source for maps and other material. Wonderfully detailed charts of the nation's coastal regions and the Great Lakes are easy to obtain and inexpensive. Not only do they tell you all you need to know about any given area—the water's depth, the characteristics of the currents, the locations of reefs and shoals, and information as to bottom conditions and the like—but they often also pinpoint the precise location of wrecked vessels.

These charts can be purchased (from $1 to $3) at local marine supply shops or map stores. Look in your Yellow Pages under "Maps."

You can also purchase charts by mail from the National Ocean Survey, an agency of the National Oceanic and Atmospheric Administration. The National Ocean Survey came into existence in 1970 when the U.S. Lake Survey was merged with the Coast and Geodetic Survey.

Some of the charts relate to long stretches of coastline while others focus upon tiny sections of harbors or other inlets. To determine the ones you need, first obtain a catalog illustrating the charts for the coastal region in which you're interested. Nautical Chart Catalog No. 1 lists the charts for the U.S. Atlantic and Gulf Coasts, including Puerto Rico and the Virgin Islands of the United States. Nautical Chart Catalog No. 2 covers

| | | | |
|---|---|---|---|
| | 11 | | Obstruction (Fish haven) |
| | Wreck showing any portion of hull or superstructure (above sounding datum) | † (Oc) | Fish haven (fishing reef) |
| | | 28 | Wreck (See O-11 to 16) |
| v 2 ft | Masts | | Wreckage    Wks |
| | 12  Wreck with only masts visible (above sounding datum) | 29 | Wreckage |
| | 13  Old symbols for wrecks | 29a | Wreck remains (dangerous only for anchoring) |
| incovers, e chart | | | Subm piles |
| | 13a Wreck always partially submerged | 30 | Submerged piling (See H-9, L-59) |
| l of chart | | | |
| nger to | †14 Sunken wreck dangerous to surface navigation (less than 11 fathoms over wreck) (See O-6a) | | *Snags    *Stumps |
| | | 30a | Snags, Submerged stumps (See L-59) |
| nown) | 5½ Wk | | |
| nger to | 15  Wreck over which depth is known | 31 | Lesser depth possible |
| | | 32 | Uncov Dries (See A-10; O-2, 10) |
| | 2½ Wk | 33 | Cov   Covers (See O-2, 10) |
| own) | 15a Wreck with depth cleared by wire drag | 34 | Uncov Uncovers (See A-10; O-2, 10) |
| nger to | | (3) | Rep (1958) |
| | 16  Sunken wreck, not dangerous to surface navigation | | Reported (with date) |
| ed rock | Foul | | Eagle Rk * (rep 1958) |

Charts available from the National Ocean Survey designate wrecks with these symbols.

the U.S. Pacific Coast, including Hawaii, Guam, and the Samoa Islands. These catalogs are free. Address your request to the National Ocean Survey, Distribution Division (C-44). Riverdale, Maryland 20840.

As for charts of the Great Lakes, these are available from the National Ocean Survey's Lake Survey Center. To obtain a chart catalog covering the Great Lakes and adjacent waterways, write to the National Ocean Survey, Lake Survey Center, 630 Federal Building, Detroit, MI 48226.

To interpret the charts, you'll need a guidebook titled *Nautical Chart Symbols and Abbreviations*. It costs 50 cents and can be obtained from any of the above sources.

Many other departments of the federal government offer information on sunken vessels. During World War II, the U.S. Naval Oceanographic Office published a series of wreck charts and a "Wreck Information List" for coastal waters of the United States. These charts have been out of print for many years. However, the Library of Congress provides photographic reproductions of sections of the "Wreck List" and specific charts. Ordering information and a rundown of prices can be obtained by writing the Library of Congress (Washington, DC 20540).

The Library of Congress has also compiled a helpful bibliography titled, *A Descriptive List of Treasure Maps and Charts*. Recently updated, the pamphlet gives background information on treasure maps, atlases, and books, and it has proven helpful to countless amateur and professional treasure seekers. Order a copy from the Superintendent of Documents (U.S. Government Printing Office, Washington, DC 20402). The pamphlet costs 45 cents.

The National Archives and Records Service (General Services Administration, Washington, DC 20408, Attention: NSL) has information relating to shipwrecks from the years 1875 to 1940. It conducts limited searches for documents relating to these sunken vessels, providing you can furnish the name of the ship and the date and place of the disaster.

The National Archives and Records Service also compiles what is known as the "Reference Service Report," a summary of wreck data covering various years and derived from many different sources, including the U.S. Coast Guard, the U.S. Merchant Marine, casualty companies, and daily newspapers. You can obtain photostat copies of the report at a cost of $1.25 for the first page and $1.00 for each additional page. For more ordering information, write to the National Archives and Records Service using the address listed above.

The U.S. Weather Bureau has published three wreck charts of the Great Lakes covering the dates 1886-1891, 1886-1893, and 1894. Photostat copies of these charts can be obtained from the National Archives and

Records Service (GSA, Cartographic Archives Division, Washington, DC 20408).

## CONDUCTING THE SEARCH

If you're wreck diving and the object of your search is a vessel that has been on the bottom more than a hundred years, don't expect to see the sharp outline of a hull or tall masts. Almost every bit of wood, except for what may be covered by silt, will have been destroyed by marine mollusks. Shipworms, specifically the teredo worm, can devour a ship's hull in a matter of a few years.

The experienced diver looks for straight lines and symmetrical shapes that indicate a man-made object of some type. Cannons, for instance. Until the mid-nineteenth century, every merchant ship was fitted out with cannons for protection. Even when covered with coral growth or lying partly buried in the sand and well rusted, the straight, tapering shapes of cannons stand out.

Most cannons are of value only as guideposts in spotting wrecks. If they are extremely old and in relatively shallow water, they may be so thoroughly oxidized that a sharp tap with a crowbar will cause disintegration. This is true only in the case of cast-iron cannons, however. Brass cannons are often well preserved. And they're valuable, each worth several thousand dollars. Cannons can be helpful in providing information about the ship that carried them. Some are engraved with the ship's name and relevant dates.

Gold is not affected by the chemical elements that make up sea water, but silver is. Through electrolytic action, silver is turned to black silver sulfide. Sometimes a silver coin is found in a perfect state of preservation next to a corroded iron spike or cannon ball. This is because the iron attracted a greater amount of electrolytic activity.

Brass objects usually sustain only slight damage, although they become thickly coated with crustations of copper salts. Pewter may become pitted but usually escapes complete destruction.

The metal that abounds in sunken vessels often serves to tip-off where the wreck lies. A magnetometer is used to ferret out the metal.

This instrument compares the intensity and direction of one magnetic field to another. It is first adjusted to zero for the normal magnetic field in the area where it is being used. When the instrument's probe passes over a site where the magnetic lines of force are distorted by the presence of iron, the metal needle begins to move. The iron below may be a cluster of old cannonballs, a cannon, or an anchor...some clue as to the location of a wreck.

In other words, the magnetometer works something like a hiker's magnetic compass. When the compass is passed over iron, the needle swings erratically.

Like so many other instruments of this general type, the magnetometer was developed during World War II. It was used to locate submerged submarines and was often suspended from a hovering blimp. Called MAD (for Magnetic Anomaly Detector) by naval personnel, the magnetometer of the 1940s was a large and complex instrument, one that required several electronic technicians to operate it.

Today's magnetometers are portable—some types can be hand-held, in fact—and are almost as easy to operate as the metal detectors you see at your local beaches. They consist of two basic components: the probe, which usually takes the form of a slim metal cylinder about the size of a rolled-up umbrella, and a control console linked to the probe by a submarine cable.

The probe contains two magnetic-sensitive "feeler" coils and is built to withstand water pressure at great depths. When towed along the sea bottom the instruments are unaffected by sea water, coral, or nonmagnetic metal. But iron, even deeply buried iron or small bits and pieces strewn about the sea floor, registers abruptly and clearly on the meter needle. One treasure seeker reported getting a magnetometer reading when the probe passed within 23 feet of a rusted iron cannon.

A treasure hunter by the name of Calvin Deviney achieved legendary success during the 1960s by virtue of his investigations of Great Lakes wrecks, and the essential piece of equipment he used was a nautical magnetometer. Housed in a torpedo-shaped container, Deviney towed the magnetometer behind his search boat. It reported the presence of ferrous metal by means of an inked stylus that marked its findings on moving graph paper.

Deviney would begin operations by laying out a search area using an automatic pilot and radio direction finder. Four main buoys were used to mark the corners of the search area, which was usually rectangular in shape. Smaller buoys were placed along the outer edges of the rectangle, each one a carefully measured distance from the one next to it. Then the boat, pulling the magnetometer behind, would make runs from one buoy to the one opposite, first covering the course in an east and west direction, then north and south. In other words, a checkerboard pattern.

Whenever the magnetometer registered a "strike," scuba divers with hand-held metal detectors were sent below to probe the site further. Deviney also used underwater television as an aid in search and recovery work.

The modern underwater metal detector. *(J. W. Fishers Mfg. Co.)*

Deviney compiled a file that listed the locations of 8,000 sunken vessels. He carefully investigated 200 of these. His divers brought to the surface guns, cannons, anchors, rudders, compasses, pottery, and an array of other artifacts.

It is mainly professional divers and treasure seekers that utilize magnetometers. For the amateur, the instrument presents two problems. Cost is one. Hand-held magnetometers begin at $500 and range up to $1,000. Sophisticated models, those that can be towed behind a boat or probe extremely deep water, can be priced from $30,000 to $50,000. The second problem is that they detect ferrous metals only. Not gold coins, in other words. What's the solution? In both cases it can be the underwater metal detector.

## UNDERWATER DETECTORS

Several companies manufacture electronic detectors for underwater use. Constructed of heavy duty plastic, these instruments are frequently about the same size and shape as a bullhorn. Grasping the pistol grip in one hand, the diver holds the unit out ahead of his body as he glides along, being careful to keep the search loop parallel to the bottom. He gets audio signals through earphones, although some units transmit responses by means of a meter or a flashing light.

Frequently these units are of the BFO type; there is a constant beep-beep sound. Underwater detectors are about as easy to operate as the machine you would use when coinshooting in a park.

These are the principal firms that manufacture and sell underwater detectors:

J. W. Fishers Manufacturing Co., Inc. (Anthony St., Taunton, MA 02780).
Kovacs Electronics (10123 Stonehurst Ave., Sun Valley, CA 91352).
Roth Industries (Box 90993, World Way Postal Center, Los Angeles, CA 90009).
States Electronics Corporation (Bludworth Marine Division, 10 Adams St., Linden, NJ 07036).
White's Electronics Inc. (1011 Pleasant Valley Rd., Sweet Home, OR 97386).

Detectors meant for underwater work are often simply modified versions of units intended primarily for land use. Exceptions to this are the several models offered by J. W. Fishers Manufacturing Company. With their unique inductance-bridge circuitry, these machines are capable of detecting both ferrous and nonferrous metals, and they operate with equal efficiency in either fresh or salt water. Divers applaud them for their sensitivity and reliability.

These units feature solid epoxy search coils which vary in size from 11 inches, common to the least expensive model, to 18 inches, in the case of the deluxe unit. Each type has a single control knob and switch. Finds are reported either by a meter, an audio (earphone) system, or both. Each model is covered by a two-year warranty.

Of the four different models in the company's product line, the Mark 3 is the most popular. Able to detect a penny at 4 inches and large objects up to depths of 4 feet, this unit is powered by a 9-volt battery (Eveready 226) and utilizes a 13-inch coil. A meter mounted just forward of the cylinder-shaped control box indicates the presence of anything metal. In air, the Mark 3 weighs 5 pounds, while underwater it has a slightly negative buoyancy. It can be operated at depths of up to 200 feet. Available at dive shops, the Mark 3 is priced at $199. If you order by mail, there's a $5.25 shipping charge.

The Mark 1, at $149, is a slightly smaller unit and not quite as sensitive as the Mark 3. If you're seeking maximum sensitivity—a penny at 4½ inches, large objects at 5½ feet—consider the Mark 5. It boasts all of the features of the Mark 3 but it has an 18-inch search coil, and an audio system as well as a meter, which enables the machine to be used in murky water and low visibility. The earphone that's provided is sealed within a waterproof case. The Mark 5 is $285; the shipping charge is $8.25.

Fishers' Mark 7 is designed to be towed from a boat; it combs the bottom for metal in big masses, a cannon, an anchor, or an outboard motor, for example. Its 32-inch coil is able to detect objects at depths of 8½ feet. The unit weighs 16 pounds.

The coil, instead of being a solid epoxy disk, has large openings to permit water to flow through, which reduces planing when the coil is being towed. A 150-foot cable is provided with the Mark 7. The unit operates off a heavy duty 9-volt battery (Eveready 266). The Mark 7 is priced at $395, and there's a $25 shipping charge.

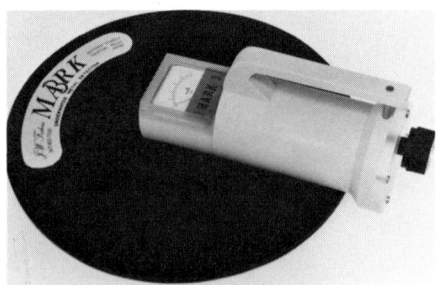 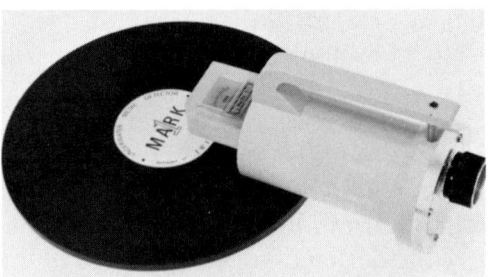

**The Mark 3.** *(J. W. Fishers Mfg. Co.)*   **The Mark 1.** *(J. W. Fishers Mfg. Co.)*

You can convert any one of these detectors to land use with the attachment of a 30-inch aluminum shaft and hand grip that the company sells. For the Mark 1 or Mark 3, the handle is $15.

Many divers have reported success using White's Goldmaster BF-8 Amphibian, a sturdy underwater unit, which, when equipped with a land probe, can also be used for beachcombing. The Amphibian weighs 3 pounds.

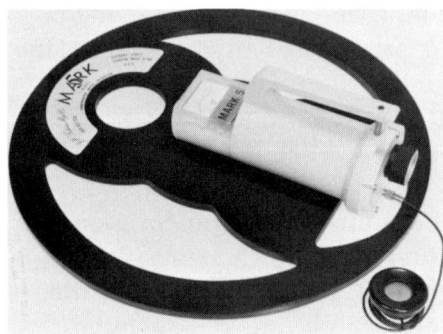 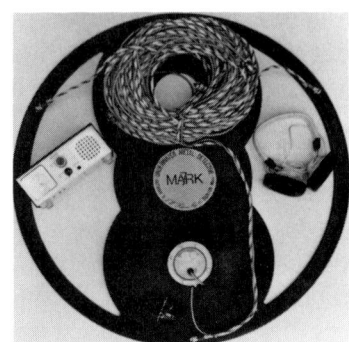

**The Mark 5.** *(J. W. Fishers Mfg. Co.)*   **The Mark 7.** *(J. W. Fishers Mfg. Co.)*

Powered by a 9-volt battery, the unit comes equipped with an 8-inch lucite search loop. It costs $199.50. To extend the depth range of the machine, you can purchase larger loops of the same type: a 12-inch loop ($59.50); 18 inches ($69.50), and 24 inches ($79.50). For seeking small objects submerged in mud, a deep water probe is available that's fitted with an 8-inch loop. It is priced at $79.50.

The Goldmaster Amphibian. *(White's Electronics)*

For beachcombing, or searching in water up to 3 feet in depth, you can obtain either a 4-inch or 6-inch coil of conventional design, plus a 42-inch handle, at $79.50. You also need an adapter, which is $10.

When using the Amphibian underwater, it's vital to stay an even distance from the bottom as you glide along. Varying the distance, particularly if the bottom happens to be high in mineral content, can cause a change in the frequency of the beats, and this can be easily misinterpreted. A diver who field-tested the Amphibian for *Skin Diver* magazine reported that he had the best results when he kept the search coil 3–4 inches from the bottom.

Water leakage is a constant problem with most submersible detectors. "It's almost impossible to keep water from getting inside the housing," says Charles Garrett, whose company was discouraged from continuing in the manufacture of such instruments by the water pressure problem.

Water, however, does not necessarily cause permanent damage. Jim White, a Portland, Oregon, diver, who has used an Amphibian, learned to cope with the water that occasionally gets inside the instrument. If the unit doesn't seem to be functioning properly during a dive, he surfaces for an inspection. He raises the unit above his head, then peers through the plexiglas search loop into the machine's interior. "Don't hold the instrument at waist level and look down into it," he warns. "You may slosh water all through the circuitry."

Any water inside should be emptied out immediately. If it happens to be salt water, flush out the unit, or at least all affected parts, with fresh water to halt corrosion. How do you dry out the machine? Jim White

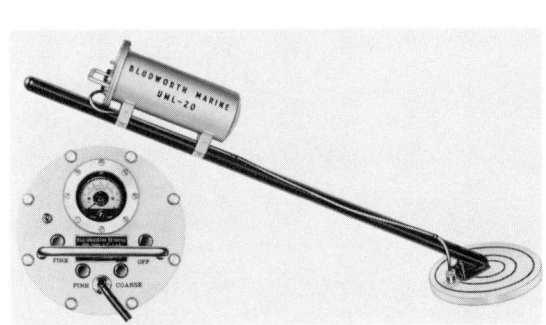

The UML-20. *(Bludworth Marine)*

suggests that, while riding in an automobile, you hold the unit outside the window. "When the Amphibian is dry," he says, "it will, with any luck, function again."

The Kovacs Company offers a compact and rugged underwater detector in the Neptune. This unit is quite unique in that it is equipped with a pair of signal lights that report detected objects. The intensity with which they shine varies directly with the strength of the signal. There's also a meter. The Neptune is equipped with a 10-inch coil and features a fine-tuning control knob. It's priced at $525.

States Electronics Corporation (Bludworth Marine Division, 10 Adams St., Linden, NJ 07036) offers a sophisticated underwater detector that is capable of perceiving any kind of metal of appreciable size in fresh or salt water, or even when buried in sand, mud, or coral. Its responsiveness is based on the conductivity of the metal and, of course, on the size of the object and how far it is from the search head. The unit is capable of detecting a silver dime buried in a foot of silt or a big piece of metal, such as a ship's anchor, at up to 4 feet.

Known as the UML-20, it can be used by a diver in up to 180 feet of water, and is pressurized to that depth, but 90 percent of its use is in depths of 50 feet or less. A meter mounted on the cylinder-shaped control housing provides a visual report on what is being detected. There is also an audio system. The instrument has been used for more than fifteen years by underwater search teams in every part of the world. It is priced at $1,475.

## THE USE OF SOUND

The sonic depth finder, sometimes known as an echo sounder, is another electronic device frequently used by modern-day treasure hunters. An instrument of this type transmits sound waves to the ocean floor which bounce back to be picked up by a receiver. The interval between the transmitted signal and the rebounded sound translates into distance covered, i.e., depth. A variation in the interval means a variation in the water's depth, a variation that can be caused by an undulation in the sea floor, a coral reef—or the hull of a sunken vessel. Some types of sonic depth finders report physical variations in the sea floor by means of a moving strip of graph paper. They can be so sensitive they will reproduce the silhouette of a ship lying on the bottom.

Scuba-Eye is the name of a recently developed hand-held instrument of this type. It's especially valuable in murky waters, such as those often encountered in Great Lakes diving. The instrument reports to the submerged diver the exact distance of the bottom from the diver's position and what type of bottom it is, that is, whether there are weeds below, scattered rocks, or coral formations—or a sunken vessel.

The sound waves sent out by the Scuba-eye rebound to the instrument and are converted into visible signals which appear on an illuminated depth meter. Different types of bottom conditions give off different types of signals. A kelp bed, for instance, produces repetitive signals, but each is of limited intensity. In the case of a solid bottom, the signal is steady and very bright. The tops of rocks or even passing fish appear as secondary signals. "There is an art to reading and interpreting the signals," said a recent article in *Skin Diver*, "but it's really quite simple to master."

Scuba-Eye can be used on objects as far distant as 200 feet. Two sturdy handles enable the diver to hold the instrument in front of his body, the way one might hold an underwater motion picture camera. The instrument's scanner is powered by 8 Type-C dry cell batteries, which deliver 12 volts.

You don't have to be an underwater diver to use Scuba-Eye. You can simply hold it over the side of the boat. It must be submerged to operate, however.

Scuba-Eye sells for $349.50. More information can be obtained from the manufacturer, Fishmaster Products (Box 9635, Tulsa, OK 74107).

## UNDERWATER SALVAGE

Salvage methods, the different techniques used in bringing what's found below to the surface, can be as varied as the methods of detection. The air lift, effective at depths of from 15 to 150 feet, is one that is used frequently.

Diving teams usually build their own air lifts because they are basically simple devices and the parts are not expensive. An air lift consists of a long metal tube or length of reinforced rubber hose, usually 6 inches in diameter. One end of the tube is suspended just above the objects to be raised, while the other end, from which the discharge will flow, is directed into a sifting system on the boat deck. Suction is created by a second hose, much smaller in diameter, that is connected to an air compressor. When compressed air is sent down the hose and introduced to the first tube at its lower end, the mixture of compressed air and water, much lower in specific gravity than the ambient water, causes a powerful sucking force at the tube mouth. The pressure is so great that an air lift can pull in and raise small cannon balls. John Potter, in his book, *The Treasure Diver's Guide*, tells of an 8-inch air lift, hovering over the cargo of the *Diamond Knot*, that sucked up 1000 gallons of water a minute and 800 tins of salmon along with it.

Sometimes the air lift is used in connection with a water jet, water from the nozzle of a 3-inch hose. Handled by divers, the jet is used to blast away sand, silt, gravel, or coral covering objects to be recovered. Once the water jet has done its work, the air lift moves in.

The age-old method of raising objects from below is to suspend a bucket or net from a rope and have a diver fill it up, whereupon the bucket is hauled aboard the boat. This principle is still being used, but instead of a bucket or net, divers often rely on inflatable pontoons. They are filled with water and sunk to the ocean floor. After being attached to the object to be raised, air is pumped into them and they rise to the surface. Divers sometimes built lifts of this type out of 53-gallon drums. John Potter says that five such drums "will raise about a ton."

Spanish coins recovered from Caribbean waters.
*(Bahamas Ministry of Tourism)*

Sometimes the bottom where the salvage work is to be performed is obscured by shifting sands. The Real 8 divers solved the problem of underwater visibility with a unique "prop wash" system. An elbow-shaped aluminum tube, 2 feet in diameter, was attached to the transom of the salvage vessel. When the boat was over the salvage site, the tube was lowered into the water and carefully positioned so that one end was just aft of the propellor. The other end pointed straight downward. Four anchors were then lowered to hold the boat in place and the engines started.

What happened was that the tube deflected the surge of water from the propellor downward. In effect, it sent clear water to the murky bottom. It also served to excavate the site, clearing away silt and coquina, the soft and porous limestone deposits that form in warm ocean waters. In 20 feet of water, the prop-wash device could excavate a hole 15 feet deep and 20 feet in diameter in 10 to 15 minutes. Once the area was excavated and the water rendered clear, divers moved in with their magnetometers and electronic detectors.

## ESTABLISHING COIN VALUES

How does one determine the value of coins that are several centuries old? It depends on many factors. If the coins are in good condition, they probably have their greatest value in the numismatic market. If they are high in gold content, it may be more worthwhile to sell them for their metallic value. It's difficult to say for sure. There's a decision to be made for each batch of coins, indeed, perhaps for each individual coin.

The chart below is meant to serve as a general guide. It gives the weight in grams for the gold coins most commonly recovered from wrecks.

*Spanish*
| | |
|---|---|
| Excellente or ducat (escudo d'oro) | 3.4 |
| Doubloon (onza, 8 escudos) | 27.0 |
| Castillano (half dobla) | 4.6 |
| Pistole (¼ doubloon or 2 escudos) | 6.5 |
| Escudo (⅛ doubloon) | 3.2 |
| Doblon de Isabela, 100 reales | 7.5 |

*French*
| | |
|---|---|
| Louis d'or | 7.2 |

*English*
| | |
|---|---|
| Guinea or sovereign (21 shillings) | 8.4 |

*Dutch*
| | |
|---|---|
| Gulden cavalier or gold rider | 10.0 |

*Portuguese*
| | |
|---|---|
| Moidore | 3.8 |
| Dobla | 4.6 |

*Brazilian*
| | |
|---|---|
| Cruzado 400 reis | 4.0 |
| Johannes | 13.0 |

It is interesting to note that the Spanish doubloon was called the onza—for ounce. But it usually did not weigh a full ounce. It was 27 grams (an ounce is approximately 31 grams).

As for silver coins—the ducato, ecu, crown, peso, piastre, eight reales (the legendary piece of eight), and duro—they usually weighed 27 grams. But the silver content varied over the years.

Never attempt to clean such coins yourself. Don't even attempt to remove coral encrustations. Consult an expert as to how they should be handled.

## TREASURE DIVING AND THE LAW

In some states, the unauthorized removal of coins or other artifacts from a shipwreck site is now considered a misdemeanor and can be punishable by a fine or imprisonment or both. Be sure you're aware of what statutes apply before you do any diving.

"It's not merely relics of obvious value that states are seeking to protect. "Shipwrecks are of concern because of the historical position they occupy in our resources and not because of their monetary nature," says a spokesman for Florida's Department of State. "We are just as interested in the wooden remains of these wrecks as we are in the coins and bars they may contain; our interest is in the total story they tell, each representing a capsule of history contained within the larger story of Spanish colonization in the New World and *LaFlorida.*

What the state of Florida does is to enter into a contract with the individual treasure seeker or professional salvage company. Under its terms, all data relating to the wreck and everything recovered is turned over to the state. After being cleaned and cataloged, 75 percent of what has been found is returned to the salvagers. Treasure seekers, individuals or groups, must first show that they have the financial capability to complete the planned research. Too many have gone bankrupt or dissolved their business because they were engulfed, not in sea water but red ink.

The state of Florida, by keeping only one-fourth of recovered treasure for itself, has to be applauded for its altruism. Indeed, state officials have been criticized for being *too* generous. The attitude of other states is quite different. Ask Paul and Max Zinica, a pair of amateur treasure hunters from Gary, Indiana. They were inspired by an article in *Argosy* magazine that told of a fleet of Spanish galleons sunk off the coast of Texas in 1553. After researching old maps, they bought metal detectors and started to search the beaches of Padre Island, a long sliver of land that lies off the southeast Texas coast. At a point just a few miles south of Corpus Christi, they began turning up silver coins.

Now they became much more serious. Supported by money from a group of Gary businessmen, the two brothers, four skin divers, and an electronics expert launched an all-out effort to find the source of the coins. Within a few months they had located several wrecks and had salvaged more than $600,000 in coins and artifacts. But the state of Texas moved in to shatter their dreams, confiscating all that the young men recovered. They are now seeking to win a court ruling that will return them their fortune, or at least a good-sized chunk of it.

To find out the policies that apply, write to the state Attorney General. In the case of Florida, you can obtain additional information from the Department of State (The Capitol, Tallahassee, FL 32304).

# NOTICE

THE LIMITED NUMBER OF HISTORICAL PERIOD SHIPWRECK SITES IN FLORIDA WATERS CONSTITUTE A SCIENTIFICALLY AND HISTORICALLY VALUABLE AND IRREPLACEABLE RESOURCE PROTECTED BY THE FLORIDA ARCHIVES AND HISTORY ACT, CHAPTER 267, FLORIDA STATUTES.

UNAUTHORIZED REMOVAL OR POSSESSION OF ARTIFACTS FROM OR DISTURBANCE OF SHIPWRECK SITES CONSTITUTES A MISDEMEANOR PUNISHABLE BY UP TO ONE YEAR IMPRISONMENT OR $500.00 FINE AND SEIZURE OF ALL BOATS, EQUIPMENT AND INSTRUMENTS USED IN CONNECTION WITH THE VIOLATION.

RULE 1A — 0.01 OF THE ADMINISTRATIVE CODE, ADOPTED PURSUANT TO PROVISIONS IN CHAPTER 267, FLORIDA STATUTES, WHICH BECAME EFFECTIVE OCTOBER 30, 1970 PROVIDES AS FOLLOWS:

Penalty: Unauthorized exploration and salvage.—

(A) Whoever removes, alters, or in any way tampers with treasure trove, artifacts, or any of the articles protected by Chapter 267, Florida Statutes, located on state lands without benefit of an exploration or salvage contract or other permit from the Division of Archives, History and Records Management shall be guilty of a misdemeanor.

(B) Whoever buys, receives, aids in the concealment of, or has in his possession such treasure trove, artifacts or other protected objects removed from state lands without benefit of a contract or permit from the Division with the knowledge that same was removed without benefit of a contract shall be guilty of a misdemeanor.

(C) In all cases of arrest and conviction under either of the above sections, all boats, instruments, and other equipment used in connection with such violation are hereby declared to be nuisances and shall be seized and carried before the court having jurisdiction of such offense for proper disposition thereof.

General Authority 267.031 FS. Law Implemented 267.061 FS.

**HELP PROTECT THIS IMPORTANT SEGMENT OF THE HISTORICAL HERITAGE OF THE PEOPLE OF FLORIDA**

The state of Florida issues this warning to treasure seekers.

# APPENDIX

BOOKS FOR TREASURE HUNTERS

Arnold, Oren. *Ghost Gold.* 1971. $3.95. Naylor.
Bancroft, Caroline. *Lost Gold Mines and Buried Treasures.* $1.25. Johnson, CO.
Brown, Mary L. *Gems for the Taking: Mine Your Own Treasure.* 1971. $5.95. Macmillan.
Campa, Arthur L. *Treasure of the Sangre De Cristo; Tales and Traditions of the Spanish Southwest.* 1963. $6.95. University of Oklahoma Press.
Coffman, Ferris L. *Atlas of Treasure Maps.* 1957. $10.00. Nelson.
Dobie, Frank J., editor. *Legends of Texas.* 1964 (reprint of 1924 edition). $7.50. Folklore.
Dwyer, John N. *Summer Gold: A Camper's Guide to Gold Panning and Metal Prospecting.* 1971. $1.95. North Star.
Eberhart, Perry. *Treasure Tales of the Rockies.* 1968. $7.00. Swallow.
Ely, Sims. *Lost Dutchman Mine.* 1953. $5.00. Morrow.
Evans, A. T. *Treasure Hunter's Yearbook.* Annual. $4.00. Eureka Press.
Gardner, Erle S. *Hunting Lost Mines by Helicopter.* 1965. $7.50. Morrow.
Garrett, Charles. *Successful Coin Hunting.* 1974. $4.95. Ram Publishing.
Horner, David L. *Treasure Galleons: Clues to Millions in Sunken Gold and Treasure.* 1971. $10.00. Dodd, Mead.
Hult, Ruby. *Treasure Hunting in the Pacific Northwest.* 1970. $5.95. Binfords.
Marx, Robert F. *Shipwrecks of the Western Hemisphere, 1492-1825.* 1971. $15.00. World.
Mitchell, John D. *Lost Mines & Hidden Treasures of the Old Frontier.* $7.50. Rio Grande.
――― *Lost Mines of the Great Southwest.* $7.50. Rio Grande.
Oles, Floyd & Helga. *Eastern Gem Trails.* $2.00. Gembooks.
Pearl, Richard M. *Colorado Gem Trails and Mineral Guide.* 1972. $5.00. Swallow.
――― *Successful Mineral Collecting and Prospecting.* $2.95. New American Library.
Rieseberg, Harry E. *Complete Guide to Buried Treasure Land & Sea.* 1969. $5.95. Fell.
Rieseberg, Harry E. & Mikalow, A. A. *Fell's Guide to Sunken Treasure Ships of the World.* 1969. $6.95. Fell.
Von Mueller, Karl. *The Treasure Hunter's Manual.* 1972. $6.00. Eight States Associates.

Wagner, Kip & Taylor, L. B., Jr. *Pieces of Eight: Recovering the Riches of a Lost Spanish Treasure Fleet.* 1966. $7.95. Dutton.
Weight, Harold O. *Lost Mines of Death Valley.* 1970. $2.50. Calico Press.
——— *Lost Mines of Old Arizona.* 1959. $2.00. Calico Press.
Wilkes, Bill S. *Nautical Archaeology: A Handbook for Skindivers.* 1971. $7.95. Stein & Day.
Zeitner, June Culp. *Midwest Gem Trails.* $2.00. Gembooks.

## TREASURE-HUNTING MAGAZINES

*Canadian Treasure* (P.O. Box 2071, Vancouver 3, B.C.). One year, 4 issues, $3.50.
*Research & Recovery* (P.O. Box 70025, Houston, TX 77007). One year, 4 issues, $3.50.
*Treasure* (Jess Publishing Co., Inc., 7950 Deering Ave., Canoga Park, CA 91304). One year, 12 issues, $12.
*Treasure Hunter* (Box 188, Midway City, CA 92655). One year, 4 issues, $3.
*Treasure Hunter's Newsletter* (Eight State Associates, Box 1438, 1918 Pearl St., Boulder, CO 80302). One year, 4 issues, $5.
*Treasure Hunting Unlimited* (422 Broadway, Truth Or Consequences, NM 87901). One year, 4 issues, $5.
*Treasure News* (P.O. Box 907, Bellflower, CA). One year, 12 issues, $3.
*Treasure Search* (Jess Publishing Co., Inc., 7950 Deering Ave., Canoga Park, CA 91304). One year, 6 issues, $6.
*Treasure World* (Drawer L, Conroe, TX 77301). One year, 6 issues, $3.50.
*True Treasure* (Drawer L, Conroe, TX 77301). One year, 6 issues, $3.50

## TREASURE-HUNTING CLUBS

*Arizona*

Arizona Treasure Hunters
2207 East Alta Vista Road
Phoenix, AZ 85040

Arizona Treasures Unlimited, Inc.
P.O. Box 14853
Phoenix, AZ 85031

*California*

Chino Valley Prospectors Club
P.O. Box 521
Chino, CA 91710

Gem & Treasure Hunting Association
3928 Twiggs St.
San Diego, CA 92110

Golden State Treasure Hounds
2209 Pageant St.
Bakersfield, CA 93396

International Explorers Club
P.O. Box 116
Paramount, CA 90723

Living West
7512 Sausalito Ave.
Canoga Park, CA 91304

Prospectors Club of Southern California
P.O. Box 43
Lomita, CA 90717

Treasure Hunters Club of Central California
2515 Mallard Drive
Walnut Creek, CA 94596

United Prospectors
5665 Park Crest Drive
San Jose, CA 95118

## Colorado

Eureka
4390 E. Mississippi Ave.
Denver, CO 80222

Mile High Gem Guide
Box 3302
High Mar Station
Boulder, CO 80303

## Florida

Bay Area Treasure Club
Box 826
Lutz, FL 33549

Peninsular Archaeological Society, Inc.
1412 Sixth Ave.
Tarpon Springs, FL 33589

Sun Coast Historical Treasure Hunters
P.O. Box 1846
Bradenton, FL 33505

## Georgia

North Georgia Relic Hunters
620 Tom Read Drive
Marietta, GA 30060

## Illinois

Illinois & Iowa Treasure Hunters
835 South Spring St.
Genesco, IL 61254

Jolly Roger Club
Wood Rd.
Ashton, IL 61006

Leif Ericson Society
Box 112
Evanston, IL 60204

Prairie Pirates Treasure Hunters Club
1009 Tonti Circle
Peoria, IL 61605

Tri-State Treasure Hunters
505 South Eighth St.
Quincy, IL 62301

## Indiana

Midwest Treasure Finders
3329 Clover Dr.
Indianapolis, IN 46231

Treasure Hunting Club of America
P.O. Box 686
Greenfield, IN 46140

## Iowa

Cedar Valley Treasure Hunters
3748 H Ave., NE
Cedar Rapids, IA 52402

## Kentucky

Kentucky Treasure Hunters
  Association
Benton, KY 42025

## Maryland

Maryland Freestate Treasure Club
6617 Loch Hill Rd.
Baltimore, MD 21239

## Massachusetts

New England Treasure Finders Club
37 Spring St.
West Springfield, MA 01089

South Shore Treasure Club
Box 313
Quincy, MA 02169

## Michigan

Michigan Treasure Hunters
610 East Wisconsin St.
Mt. Pleasant, MI 48858

Tri-City Treasure Hunters
4512 Quincy Dr.
Midland, MI 48640

## Missouri

Adventurer's Club
Box 244
St. Charles, MO 63301

*Nebraska*

Nebraska Treasure Hunter
Sprague, NE 68438

*New York*

Atlantic Treasure Club
1021 Park Blvd.
Massapequa Park, NY 11762

Bronx Explorers
Box 868
Bronx, NY 10402

New York State Metal Detector Club
366 Brookview Dr.
Rochester, NY 14617

Treasure Trove Club
2922 174th St.
Flushing, NY 11385

*North Carolina*

South Eastern Association
  of Relic and Coin Hunters
123 East Davis St.
Burlington, NC 27215

*Ohio*

American Treasure Hunters Society
10326 Lorain Ave.
Cleveland, OH 44111

Buckeye Treasure Hunters Club
218 N. Market St.
Minerva, OH 44657

Detector Club of Northern Ohio
Park Blvd.
4130

*Oklahoma*

Greene County Treasure Club
1588 Colorado Ave.
Bartlesville, OK 74003

Indian Territory Treasure Hunters
4972 East 26th St.
Tulsa, OK 74134

*Oregon*

Albany Metal Detectors Club
P.O. Box 203
Albany, OR 97321

Oregon Treasure Hunters League
P.O. Box 14052
Portland, OR 97214

*Texas*

Houston Treasure & Relic Hunters
726 Redan St.
Houston, TX 77009

Prospectors Club of Odessa
P.O. Box 1215
Odessa, TX 79760

Relic & Treasure Club of Austin
4804 Gladview Drive
Austin, TX 78745

Research & Recovery Club of
  Pasadena
1213 Preston Rd.
Pasadena, TX 77503

Southern Texas Treasure Hunting
  Club
350 Edgebrook Lane
San Antonio, TX 78213

Treasure Hunters Association of
  Pasadena
415 Shoreacres Blvd.
La Porte, TX 77571

*Washington*

Northwest Ghost Towners Coin &
  Treasure Club
24615 NE. 18th St.
Redmond, WA 98052

Northwest Treasure Hunters
1112 East Hoffman Ave.
Spokane, WA 99207

*Wisconsin*

Wisconsin Association of Treasure
  Hunters
13327 W. Greenfield Ave.
New Berlin, WI 53151

## STATES OFFERING INFORMATION ON PROSPECTING AND MINERAL COLLECTING

*Where to write:*      *What to write for:*

### Alabama

Publication Sales
Geological Survey of Alabama
P.O. Drawer O
University, AL 35486

*Rocks and Minerals of Alabama; A Guidebook for Rockhounds.* A comprehensive guide prepared by the Alabama Geological Survey. Cost, $1.

### Alaska

State of Alaska
Department of Natural Resources
Division of Geological
   and Geophysical Surveys
P.O. Box 80007
College, AK 99701

Annual Report, 1972; Division of Geological and Geophysical Surveys. A 64-page review of earth study programs, plus a comprehensive list of available information circulars.

State of Alaska
Department of Natural Resources
Division of Mines and Geology
Box 5-300
College, AK 99701

"Information Circular No. 3, Hand Placer Mining Methods"; "Information Circular No. 6, Alaskan Prospecting Information"; and "Information Circular No. 18, Amateur Gold Prospecting in Alaska."

### Arizona

Arizona Bureau of Mines
The University of Arizona
Tucson, AZ 85721

"List of Available Publications." A listing of maps and bulletins, plus a rundown of services provided by the Bureau of mines.

Arizona Department of Mineral
   Resources
Mineral Building
Fair Grounds
Phoenix, AZ 85007

Bulletins entitled, "Director of Rockhound Clubs in America," "Books for Mineral Collectors," and "Arizona Dealers in Mineral Specimens."

State of Arizona
Department of Economic Planning
   and Development
3003 North Central Ave.
Phoenix, AZ 85012

"Arizona Rockhound Guide." An introductory brochure describing mineral locations by county.

*Rockhound Primer of Arizona.* A 64-page brochure, complete and colorful, explaining and picturing the state's rocks and minerals, and telling where to find them. Highly recommended.

### Arkansas

Arkansas Geological Commission
Little Rock, AR 72204

"Rock and Mineral Collecting Localities." A brochure containing maps and descriptive information.

*Where to write:*

## California

California Division of Mines
  and Geology
Publications Sales
P.O. Box 2980
Sacramento, CA 95812

*What to write for:*

"Gold Packet"—A kit of materials that includes bulletins titled, "Tips for Gold Hunters," "Lands Open to Prospecting," "Marketing of Gold," "Gold Refineries," "Index of Geologic Atlas," "Mining Legal Forms," "Assessment Work on Mining Claims," and many others. Kit cost, $1.

*List of Available Publications.* A 32-page brochure listing bulletins, reports, and various types of maps available from the Division of Mines and Geology.

*California Geology.* A monthly magazine published by the Division of Mines and Geology whose purpose is "to report on the progress of earth science in California and inform the public of discoveries in geology and allied sciences." Subscription rate, $2 a year; a bargain.

*Basic Placer Mining.* Informative 16-page bulletin. Cost, 10 cents.

## Colorado

Colorado Bureau of Mines
215 Columbine Building
1845 Sherman St.
Denver, CO 80203

Bulletins titled, "Placer Mining," "Common Ores of Colorado," and "Colorado Gemstones."

## Connecticut

State of Connecticut
Connecticut State Library
231 Capitol Ave.
Hartford, CT 06115

Bulletin of general information to rockhounds.

## Delaware

State of Delaware
Division of Economic Development
45 The Green
Dover, DE 19901

Circulars entitled: "Generalized Geologic Map of Delaware," "Geologic History of Delaware," and "Selecting Fossil-Collecting Locations in Delaware."

*Where to write:*

## Idaho

Caxton Printers
Caldwell, ID 83605

Department of Commerce &
  Development
State of Idaho
Room 108
Capitol Bldg.
Boise, ID 83707

## Indiana

State of Indiana
Department of Natural Resources
Geological Survey
611 North Walnut Grove
Bloomington, IN 47401

## Iowa

Iowa Conservation Commission
300 Fourth St.
Des Moines, IA 50319

## Kansas

Kansas Geological Survey
The University of Kansas
1930 Avenue A
Campus West
Lawrence, KS 66044

## Kentucky

Kentucky Geological Survey
307 Mineral Industries Building
University of Kentucky
Lexington, KY 40506

*What to write for:*

*Gem Minerals of Idaho,* by John A. Beckwith. A recently published paperback, this book features a chapter on each of the gem minerals known to occur in the state and eleven maps outlining field trips.

"Gems and Locations." Small Brochure.

*Minerals of Indiana.* A 74-page handbook, complete and authoritative. A "must" item. Cost, 75¢.

"Gold and Diamonds in Indiana." A reprint of a report first published in 1903 describing the distribution of gold and diamonds. Cost, 35¢.

"Publications of General Interest." A bulletin listing circulars, reports, and maps available to the general public, with the price for each.

A 6-page excerpt from *Outdoor Recreation In Iowa* which describes the state's geological formations.

*Kansas Rocks and Minerals.* A colorful, nontechnical 64-page booklet that summarizes rock and mineral occurrences. Excellent.

*Kentucky's Rocks and Minerals.* A scientific guide, written in layman's terms, to 40 rocks and minerals that occur in the state; also general information on collecting. Highly recommended. Cost, $1.

*Where to write:*

*What to write for:*

"List of Publications." A 42-page booklet; geologic maps are among the listings.

## Maine

State of Maine
Department of Conservation
State Office Building
Augusta, ME 04330

"Publications of the Maine Geological Survey." A 4-page bulletin.

## Massachusetts

The Museum of Science
236 State St.
Springfield, MA 01103

A booklet describing minerals and rocks of Springfield and vicinity. Cost, 65¢.

The Museum Shop
Museum of Comparative Zoology
24 Oxford St.
Harvard University
Cambridge, MA 02138

A booklet describing Massachusetts mineral localities. Cost, 50¢.

## Michigan

State of Michigan Tourist Council
Department of Natural Resources
Commerce Center Building
300 S. Capitol Ave.
Lansing, MI 48926

*Collecting Minerals in Michigan.* Excellent general study; 26 pages.
   "Geological Survey Pamphlet No. 3, Guide to Michigan Fossils."
   "Michigan's Colorful Minerals." A 4-page brochure.
   "Michigan's Beach Stones." A 4-page brochure.

## Minnesota

State of Minnesota
Department of Economic
   Development
51 East 8th St.
St. Paul, MN 55101

Bulletin titled, "Minnesota Rocks, Minerals, Fossils and Where to Find Them."

## Mississippi

Mississippi Geological Economic
   and Topographical Survey
2525 N. West St.
P.O. Box 4915
Jackson, MS 39216

*Fossil and Mineral Collecting Localities of Mississippi.* County-by-county maps and descriptive information; excellent. 75¢.
   "List of Publications." Descriptive information and prices for official bulletins and maps; 21 pages.

*Where to write:*

## Missouri

Missouri Geological Survey and
 Water Resources
Buehler Park
Rolla, MO 65401

## Montana

Advertising Unit
Montana Department of Highways
Helena, MT 59601

Union Bank and Trust Company
Helena, MT

## Nevada

Department of Economic
 Development
Capitol Building
Carson City, NV 89701

## New Hampshire

State of New Hampshire
Division of Economic Development
Concord, NH 03301

## New Jersey

Map & Publication Sales Office
Bureau of Geology and Topography
Trenton, NJ 08625

## New Mexico

Bureau of Land Management
Albuquerque District Office
3550 Pan American Freeway, NE
Albuquerque, NM 87107

State of New Mexico
Bureau of Mines and Mineral
 Resources
Socorro, NM 87801

*What to write for:*

A list of publications; a list of the 17 Missouri rockhound clubs.

Bulletin of general descriptive information concerning precious, semiprecious, and unusual stones and their localities.

Brochure titled, "So You Want to Pan Gold."

"Nevada Rock Hunting," brochure containing general descriptive information.

Brochure titled, "Some Pointers on Mineral Collecting in New Hampshire."

"Publications." A brochure listing reports, bulletins, maps, pamphlets, and miscellaneous publications available from the Bureau of Geology and Topography.

"A guide to Public Lands in New Mexico." A large, detailed, and colorful map.

"Bulletin No. 87." A summary report on gem minerals and their localities, plus bibliographic references; 437 pages. Cost, $2.50.
 Bulletin describing the geologic and mineralogic features of Rock Hound State Park.

*Where to write:*

### New York

University of the State of New York
The State Education Department
New York State Museum and
 Science Service
Albany, NY 12224

*What to write for:*

Pamphlet titled, "Gems of New York State."
 "Generalized Geological Map of New York State."
 "A Complete List of Geological Publications." A rundown on the bulletins, handbooks, circulars, and maps available from the State Museum and Science Service.

### North Carolina

State of North Carolina
Department of Natural Resources
Box 27687
Raleigh, NC 27611

Information Circular 16, "Mineral Localities of North Carolina." Cost, $1.
 Information Circular 21, "The Gold Resources of North Carolina." Cost, $1.
 Educational Series 1, "An Introduction to the Topography, Geology, and Mineral Resources of North Carolina."
 "List of Publications of the Division of Mineral Resources."

State of North Carolina
Travel and Promotion Division
Department of Natural and
 Economic Resources
Raleigh, NC 27611

"Gems of North Carolina." Brochure of general information.

### North Dakota

North Dakota Highway Department
Capitol Grounds
Bismarck, ND 58501

"Roughrider Guide to North Dakota." Travel and promotion booklet with 2-page section on minerals and fossils.

### Ohio

Ohio Department of Natural
 Resources
Division of Geological Survey
Fountain Square
Columbus, OH 43224

"List of Publications." Comprehensive rundown on data available concerning the state's mineralogical and geological resources.
 "Flint; Ohio's Official Gemstone." Colorful pamphlet on the most popular of the state's minerals.

*Where to write:*

## Oregon

State of Oregon
Department of Geology
 and Mineral Industries
1069 State Office Building
Portland, OR 97201

*What to write for:*

"Publications List." A 4-page bulletin.
"Thunder Egg: Oregon's State Rock." Interesting brochure on thunder eggs, how they form, where they occur.
"Oregon Rocks, Fossils & Minerals." Pamphlet giving general information on where to find rocks and minerals; also lists rockhound clubs.
"Central Oregon Rockhound Guide." Colorful map and guide of Prineville area.

## Pennsylvania

Commonwealth of Pennsylvania
Department of Environmental
 Resources
Topographic and Geologic Survey
Harrisburg, PA 17120

*Mineral Collecting in Pennsylvania.* A comprehensive 164-page guide describing mineral localities; also covers minerals and mineral collecting. Excellent. Cost, 50¢.
*Pennsylvania Geological Publications.* a 64-page booklet; many of the publications are indexed by county.

## Rhode Island

Rhode Island Development Council
Roger Williams Building
Hayes St.
Providence, RI 02908

Bulletins titled, "Reference Material for Rockhounds," "General Geology," and "The Rhode Island Landscape, Underlying Rocks."

## South Carolina

State of South Carolina
State Development Board
Division of Geology
P.O. Box 927
Columbia, SC 29202

"Geologic Publications." A listing of bulletins, reports, and maps available from the Division of Geology.

## South Dakota

Department of Natural Resource
 Development
South Dakota Geological Survey
Science Center
University of South Dakota
Vermillion, SD 57069

"Publications of the South Dakota Geological Survey."

*Where to write:*

### Tennessee

Tennessee Department of
  Conservation
Division of Geology
G-5 State Office Building
Nashville, TN 37219

### Texas

American Association of
  Petroleum Geologists
P.O. Box 979
Tulsa, OK 74101

Bureau of Economic Geology
The University of Texas at Austin
Austin, TX 78712

### Utah

State of Utah
Utah Geological and Mineral
  Survey
103 UGS Building
University of Utah
Salt Lake City, UT 84112

### Vermont

State of Vermont
Department of Libraries
Montpelier, VT 05602

### Virginia

Commonwealth of Virginia
Department of Conservation
  and Economic Development
Division of Mineral Resources
Box 3667
Charlottesville, VA 22903

*What to write for:*

"List of Publications." A comprehensive rundown on the bulletins, reports, maps, and guidebooks concerning the state's topography, geology, and mineral resources.

"Geological Highway Map of Texas." Cost, $2.50.

"Texas Rocks and Minerals; An Amateur's Guide" (Guidebook No. 6), and "Texas Gemstones" (Report of Investigations No. 42). General locality information prepared by the Bureau of Economic Geology.

Bulletins titled, "Gem Localities of Utah" and "Utah Fossil Localities."

"Publications of the Vermont Geological Survey"; a 4-page bulletin.

"Mineral and Fossil-Collecting localities in Virginia." An 8-page reprint of an article that appeared in the February 1974, issue of *Virginia Minerals;* complete, authoritative. Cost, 25¢.

"List of Publications." A listing of maps and publications relating to the state's geology and mineral resources.

*Where to write:*

## Washington

State of Washington
Division of Geology and
  Earth Resources
Department of Natural
  Resources
Olympia, WA 98504

*What to write for:*

Bulletins titled, "Rocks, Fossils, and Minerals in Washington"; "Localities of Gems and Ornamental Stones"; "Names and Addresses of Mineral Societies in Washington"; and "Selected Bibliography on Gemstones."

"Geologic Publications of the Geology and Earth Resources Division." A listing of bulletins, circulars, reprints, reports, and maps available by mail.

## Wisconsin

University of Wisconsin—
  Extension
Geological and Natural
  History Survey
1815 University Ave.
Madison, WI 53706

"Mineral and Rock Collecting in Wisconsin." A 12-page survey covering types of rocks and minerals to be found in Wisconsin, plus a brief geologic history of the state. Very good.

"Publications of the University of Wisconsin Geological and Natural History Survey." A 22-page booklet.

## Wyoming

The Geological Survey of Wyoming
Box 3008
University of Wyoming
University Station
Laramie, WY 82071

"Publications Available from the Geological Survey of Wyoming." Listing of maps and reports.

## LOCALITIES IN THE UNITED STATES WHERE GEM MATERIALS ARE FOUND

*State; County*          *Gem Materials*

### Alabama:
Clay   Azurite, garnet, malachite, marcasite.
Cleburne   Beryl, garnet, kyanite, malachite, marcasite.
Coosa.   Beryl, tourmaline.
Limestone   Geodes
Tuscaloosa   Agate, chert, jasper, onyx.

### Alaska (by region):
Cook Inlet-Susitna   Soapstone.
Northwestern Alaska   Jade.
Southeastern Alaska   Agate, garnet, jade.

*State; County*          *Gem Materials*

### Arizona:
Apache   Agate, obsidian, peridot, petrified wood.
Cochise   Agate, amethyst, azurite, malachite, shattuckite.
Coconino   Agate, obsidian, petrified wood.
Gila   Agate, amethyst, azurite, bloodstone, carnelian, chalcedony, chrysocolla, epidote, garnet, hypersthene, obsidian, opal, peridot, petrified wood, serpentine, turquoise.

| State; County | Gem Materials |
| --- | --- |

Graham  Agate, azurite, banded agate, chalcedony, malachite, obsidian, onyx, opal, petrified wood, turquoise.
Greenlee  Agate, amethyst, azurite, chalcedony, chrysocolla, garnet, jasper, malachite, obsidian, onyx, opal, petrified wood, rose quartz, shattuckite, turquoise, variscite.
Maricopa  Agate, amethyst, chalcedony, fire agate, jasper, marble, onyx, opal, petrified wood, plume agate.
Mohave  Agate, chalcedony, chalk turquoise, jasper, petrified wood.
Navajo  Garnet, petrified wood.
Pima  Azurite, chalcedony, malachite, obsidian, shattuckite.
Pinal  Chalcedony, chrysocolla, obsidian, peridot.
Yavapai  Agate, carnelian, chalcedony, chrysocolla, chrysoprase, jade, jasper, moonstone, obsidian, onyx, petrified wood, quartz crystal.
Yuma  Agate, chalcedony, garnet, jasper, obsidian, petrified wood, quartz crystal, rhyolite, turquoise.

## *Arkansas:*
Garland  Novaculite, quartz crystal.
Hot Spring  Quartz crystal, smoky quartz.
Montgomery  Quartz crystal.
Pike  Diamond.

## *California:*
Alameda  Agate.
Alpine  Agate.
Amador  Rhodonite.
Calaveras  Agate, chalcedony, chrysoprase, petrified wood, quartz crystal.
Colusa  Travertine.
Contra Costa  Agate.
Del Norte  Agate, jasper, petrified wood.
El Dorado  Agate, garnet, idocrase, jasper, nephrite, petrified wood, vesuvianite.
Fresno  Chert, jasper, petrified wood, smoky quartz.
Humboldt  Agate, jade, jasper, petrified wood.
Imperial  Agate, andalusite, chalcedony, dumortierite, garnet, jasper, kyanite, opal, petrified wood, petrified palm root.
Inyo  Agate, amethyst, bloodstone, epidote, garnet, geode, jasper, obsidian, onyx, opal, quartz crystal, turquoise.
Kern  Actinolite, agate, colemanite crystal, jade, jasper, morrisonite, petrified wood, quartz crystal, rhodonite, rose quartz.
Lake  Jasper, onyx, quartz crystal.
Los Angeles  Agate, chalcedony, jasper, rhodonite.

Madera  Chiastolites.
Marin  Agate, jade, petrified whalebone.
Mariposa  Quartz crystal.
Mendocino  Agate, jade, jasper, opal, quartz.
Modoc  Agate, jasper, obsidian.
Mono  Geode, obsidian.
Monterey  Agate, jadeite, jasper, nephrite, rhodonite, serpentine.
Napa  Cinnabar, jasper, onyx, quartz crystal.
Nevada  Opal, petrified wood.
Placer  Jade.
Riverside  Actinolite, agate, amazonite, aquamarine, beryl, calcite, chalcedony, corundum, diopside, epidote, fire agate, garnet, geode, idocrase, jasper, petrified wood, quartz crystal, rhodonite, rose quartz, rubellite, spinel, topaz, tourmaline.
Sacramento  Agate, jade, opal.
San Benito  Benitoite, diopside, garnet, jade, jadeite, natrolite, serpentine.
San Bernardino  Actinolite, agate, amethyst, aragonite, bloodstone, chalcedony, crawfordite, epidote, geode, jasper, moss agate, onyx, opalite, petrified wood, rhodonite, travertine, verde antique.
San Diego  Beryl, garnet, kunzite, quartz crystal, rhodonite, topaz, tourmaline.
San Francisco  Jasper, nephrite.
San Luis Obispo  Agate, jade, jasper, marcasite, onyx, quartz, sagenite.
San Mateo  Petrified whalebone.
Santa Barbara  Agate.
Santa Clara  Agate, jasper.
Santa Cruz  Agate, petrified whalebone.
Siskiyou  Agate, californite, idocrase, jade, quartz crystal, rhodonite.
Solano  Onyx, travertine.
Sonoma  Actinolite, jasper.
Trinity  Agate, jade, jasper, rhodonite.
Tulare  Agate, chrysoprase, jade, quartz crystal, smoky quartz, thulite, topaz.
Tuolumne  Agate, gold quartz, jasper, marble, petrified wood, pyrite.
Ventura  Agate.

## *Colorado:*
Chaffee  Agate, aquamarine, beryl, jasper, onyx, petrified wood, phenacite.
Clear Creek  Amazonite, garnet.
Custer  Petrified wood.
Delta  Jasper.
Douglas  Amazonite, smoky quartz, topaz, petrified wood.
Elbert  Petrified wood.
El Paso  Agate, amazonite, garnet, phenacite, topaz, tourmaline, zircon.
Fremont  Agate, coprolite, onyx, petrified wood, rose quartz, satin spar.

| State; County | Gem Materials |
|---|---|

Garfield  Oil shale, petrified wood.
Jefferson  Amazonite, barite crystal, beryl, topaz, tourmaline.
Kiowa  Agate.
Las Animas  Rose agate.
Mesa  Agate, amethyst, flint, petrified wood.
Mineral  Agate amethyst, jasper, petrified wood.
Moffat  Agate.
Montrose  Amazonite, coprolite, covellite, jasper, phenacite, smoky quartz, topaz.
Ouray  Quartz Crystal, rhodonite.
Park  Agate, beryl, fluorite, jade, moss opal, petrified wood, topaz.
Rio Grande  Agate, petrified wood.
Saguache  Agate, amethyst, jade, turquoise.
San Juan  Feldspar, mica, quartz crystal, rhodonite.
San Miguel  Jasper, petrified bone.
Sedgwick  Agate, petrified wood.
Teller  Agate, amazonite, amethyst, jade, petrified wood, phenacite, quartz crystalsiderite, smoky quartz, topaz, zircon.
Weld  Agate, Barite, petrified wood.

## *Connecticut:*

Fairfield  Albite, beryl, epidote, fluorite, mica, quartz crystal, scheelite, siderite, spodumene, topaz, tremolite, wolframite.
Hartford  Amethyst, azurite, datolite, malachite, phrenite, quartz, smoky quartz.
Litchfield  Calcite, fluorite, galena, garnet, graphite, kyanite, magnetite, opal, prehnite, pyrite, quartz, siderite, sphalerite, stilbite, tourmaline, tremolite.
Middlesex  Actinolite, amblygonite, apatite, austinite, bertrandite, beryl, columbite, chrysoberyl, feldspar, fluorite, garnet, hiddenite, kunzite, lepidolite, mica, pollucite, rose quartz, spodumene, topaz, tourmaline.
New Haven  Amethyst, barite, beryl, columbite, datolite, feldspar, garnet, phrenite, rose quartz, smoky quartz.
New London  Garnet, kyanite, pyrite, quartz crystal, rose quartz.

## *Florida:*
Hillsborough  Agatized coral.
Pinellas  Agatized shark bone.

## *Georgia:*
Brooks Agate.
Cobb  Topaz.
Elbert  Smoky quartz.
Forsyth  Moonstone.
Habersham  Kyanite.
Jones  Agate.

Lincoln  Lazulite, rutile.
Lumpkin  Garnet.
Morgan  Amethyst.
Paulding  Garnet.
Rabun  Amethyst, garnet, quartz, rose quartz, ruby.
Towns  Ruby, sapphire.
Troup  Amethyst, aquamarine, rose quartz.
Washington  Opal.
Wilkes  Kyanite, lazulite, rutile.

## *Idaho:*
Benewah  Garnet.
Blaine  Agate, petrified wood.
Butte  Agate, petrified wood.
Canyon  Agate, petrified wood, white plume agate.
Lemhi  Petrified wood.
Nez Perce  Agate, garnet, jasper, opal, petrified wood, sapphire.
Owyhee  Agate, jasper, petrified wood.

## *Illinois:*
Hancock  Geode.

## *Iowa:*
Des Moines  Geode.
Henry  Agate, jasper, petrified wood.
Lee  Geode, jasper.
Muscatine  Agate.
Page  Agate.

## *Kansas:*
Cherokee  Marcasite, sphalerite.
Franklin  Petrified wood.
Wallace  Opal.
Wyandotte  Agate.

## *Louisiana:*
Ouachita  Agate, jasper, petrified wood.
Vernon  Petrified wood.

## *Maine:*
Androscoggin  Amblygonite, apatite, beryl, garnet, lepidolite, spodumene, pollucite, tourmaline.
Cumberland  Actinolite, beryl, columbite, garnet.
Franklin  Columbite, spodumene.
Hancock  Beryl.
Kennebec  Cancrinite, nephelite, sodalite, zircon.
Knox  Spodumene.
Oxford  Agate, amblygonite, amethyst, apatite, aquamarine, beryl, chrysoberyl, eosphorite, herderite, lepidolite, pollucite, quartz crystal, rose quartz, smoky quartz, spodumene, tourmaline.

| State; County | Gem Materials |
|---|---|

Sagadahoc   Beryl, lepidolite, tourmaline.
Washington   Agate, jasper.

## Maryland:
Allegheny   Barite crystal. quartz crystal, siderite.
Baltimore   Antigorite, calcite crystal, garnet, jasper, quartz, serpentine, smoky quartz, tourmaline.
Calvert   Jasper.
Carroll   Azurite, malachite.
Cecil   Onyx, serpentine, williamsite.
Frederick   Onyx, malachite.
Garrett   Picrolite, serpentine, williamsite.
Hartford   Serpentine.
Washington   Cuprite.

## Massachusetts:
Hampden   Amethyst, beryl, datolite, marcasite, phrenite.
Hampshire   Amethyst, beryl, datolite, galena, phrenite, pollucite, spodumene.
Middlesex   Scheelite.
Worcester   Beryl, chiastolite, spodumene.

## Michigan:
Dickinson   Actinolite.
Emmet   Agate, fossils, petoskey stone.
Houghton   Agate, garnet.
Keweenaw   Agate, chlorastrolite, domeykite, thomsonite.
Marquette   Jasper, jaspillite.
Ontonagon   Agate, datolite, malachite, tenorite.

## Minnesota:
Carlton   Agate, jasper.
Cook   Agate, heulandite.
Lake   Agate, thomsonite.
Morrison   Garnet, staurolite.
Pipestone   Catlinite.
St. Louis   Agate.
Winona   Agate.

## Mississippi:
Harrison   Jasper.
Wayne   Petrified wood.

## Missouri:
Bollinger   Agate, jasper.
Cape Girardeau   Agate, jasper.
Clark   Geode.
Crawford   Amethyst.
Franklin   Amethyst.
Jackson   Amethyst.
Lewis   Agate.
Madison   Agate, jasper.
St. Louis   Agate, barite, galena, geode.
Wayne   Agate, jasper.

| State; County | Gem Materials |
|---|---|

## Montana:
Beaverhead   Corundum, quartz.
Custer   Agate, Sapphire.
Dawson   Agate.
Deer Lodge   Sapphire.
Fergus   Sapphire.
Gallatin   Agate, corundum, petrified wood, rose quartz.
Granite   Sapphire.
Jefferson   Amethyst, barite, tourmaline.
Judith Basin   Sapphire.
Lewis and Clark   Garnet, ruby, sapphire, spinel.
Madison   Garnet, jasper, onyx, quartz crystal, serpentine, tourmaline.
Meagher   Agate.
Missoula   Quartz.
Park   Amethyst, arsenopyrite, garnet, iceland spar, petrified wood, travertine.
Powell   Agate, amazonite sapphire.
Prairie   Agate, petrified wood.
Ravalli   Fluorite.
Rosebud   Agate.
Silver Bow   Amethyst, epidote, fluorite, garnet, sapphire.
Yellowstone   Agate, jasper.

## Nebraska:
Cass   Horn coral, petrified wood.
Dawes   Chalcedony.
Gage   Geode.
Jefferson   Agate, jasper, petrified wood.
Lincoln   Agate, jasper.
Nance   Agate, jasper, petrified wood.
Saunders   Agate, petrified wood.
Sioux   Agate, chalcedony, petrified wood.

## Nevada:
Clark   Amethyst.
Douglas   Topaz.
Elko   Agate, azurite, chalcedony, petrified wood.
Esmeralda   Agate, chalcedony, petrified wood, turquoise.
Eureka   Sulfur crystal.
Humboldt   Agate, fire opal, jasper, petrified wood, rhodonite.
Lander   Chert, turquoise.
Lincoln   Agate, chalcedony, chrysocolla, jasper, malachite, petrified wood, quartz.
Lyon   Agate, jasper, opal.
Mineral   Obsidian, petrified wood, turquoise.
Nye   Onyx, petrified wood.
Pershing   Opal, petrified wood.
Washoe   Agate, garnet, idocrase, jasper, obsidian, petrified wood, piedmontite.
White Pine   Garnet.

| *State; County* | *Gem Materials* |

### New Hampshire:
Carroll  Amethyst, danalite, helvite, phenacite, quartz crystal, smoky quartz, topaz.
Cheshire  Amethyst, apatite, aquamarine, beryl, fluorite, garnet, quartz crystal, rose quartz, tourmaline, spodumene.
Coos  Amethyst, quartz crystal, topaz.
Grafton  Apatite, beryl, columbite.
Merrimack  Beryl, garnet.
Rockingham  Spodumene.
Sullivan  Aquamarine, rose quartz.

### New Jersey:
Bergen  Natrolite.
Cape May  Jasper, quartz crystal.
Mercer  Albite, calcite, chabazite, datolite, natrolite, stilbite.
Middlesex  Marcasite, petrified wood, pyrite.
Morris  Carnelian, serpentine.
Passaic  Agate, amethyst, carnelian, chabazite, datolite, heulandite, pectolite, prehnite.
Sussex  Aragonite, corundum, garnet, pyrrhotite, rhodonite, spinel, tourmaline, willemite.
Union  Calcite, chalcedony, prehnite, sphalerite, stilbite.
Warren  Calcite, chalcedony, molybdenite, prehnite, sphalerite.

### New Mexico:
Bernalillo  Agate.
Catron  Agate.
Chaves  Agate.
Eddy  Agate, quartz crystal.
Grant  Agate.
Hidalgo  Agate, chalcedony, serpentine.
Luna  Agate, carnelian, onyx, travertine.
Otero  Turquoise.
Rio Arriba  Agate beryl, dumortierite, feldspar, petrified wood.
San Juan  Agate, petrified wood, ricolite.
Santa Fe  Beryl.
Sierra  Agate, carnelian, fluorite, petrified wood.
Socorro  Agate, chalcedony.
Valencia  Agate, jasper, obsidian, petrified wood.

### New York:
Dutchess  Quartz crystal.
Erie  Satin spar.
Essex  Garnet, rose quartz, wollastonite.
Herkimer  Quartz crystal.
Jefferson  Hematite.
Madison  Celestite.
Niagara  Calcite.
Orange  Quartz crystal, sunstone, tourmaline.
Putnam  Magnetite, opal, serpentine.
Rockland  Pink garnet, spene.
St. Lawrence  Calcite, hexagonite, pyrite, serpentine, sphalerite, talc, tremolite.
Saratoga  Beryl.
Ulster  Quartz crystal.
Warren  Garnet.
Westchester  Beryl, garnet, quartz, rose quartz.

### North Carolina:
Alexander  Hiddenite, quartz, rutile, sapphire.
Ashe  Beryl, garnet, moonstone, rutile.
Avery  Epidote, unakite.
Buncombe  Kyanite, moonstone.
Burke  Amethyst.
Clay  Ruby, sapphire.
Granville  Jasper.
Haywood  Emerald, sapphire.
Iredell  Actinolite, rose quartz, sapphire.
Macon  Garnet, rhodonite, ruby, sapphire.
Mitchell  Actinolite, beryl, biotite, emerald, epidote, feldspar gems, garnet, rhodonite, unakite.
Orange  Agate.
Rutherford  Emerald.
Warren  Amethyst.
Wilkes  Agate.
Yancey  Emerald, feldspar gems.

### North Dakota:
Adams  Petrified wood.
Billings  Petrified wood.
McLean  Petrified wood.
Morton  Petrified wood.
Stark  Chalcedony, jasper, quartz crystal.

### Ohio:
Coshocton  Flint, selenite crystal.
Franklin  Petrified wood.
Gallia  Petrified wood.
Licking  Flint, jasper.
Lucas  Fossils.
Montgomery  Agate, gem granite.
Muskingum  Jasper.
Ottawa  Celestite, fluorite.
Wood  Barite, calcite.

### Oklahoma:
Canadian  Agate, jasper, petrified wood.
Comanche  Zircon.
Dewey  Agate, chalcedony, jadite, jasper, petrified wood.
Greer  Alabaster.
Jackson  Quartz.
Major  Agate, jasper, petrified wood.
Ottawa  Sphalerite.
Pushmataha  Green quartz.

| State; County | Gem Materials |
|---|---|

### Oregon:
Baker  Agate, jasper, petrified wood.
Benton  Agate.
Coos  Fossil wood.
Crook  Agate, carnelian, geode, moss agate.
Curry  Jade.
Deschutes  Agate, carnelian, geode, jasper, moss agate.
Douglas  Agate.
Grant  Agate, petrified wood.
Harney  Agate, obsidian.
Jackson  Agate, bloodstone, jasper, petrified wood, rhodonite.
Jefferson  Agate, amethyst, geode, opal.
Lake  Geode, obsidian.
Lane  Agate, petrified wood.
Lincoln  Agate, agatized coral, bloodstone, jasper, petrified wood, sagenite, sardonyx.
Linn  Agate.
Malheur  Agate, geode, jasper, petrified wood.
Morrow  Agate, geode.
Polk  Agate, jasper, petrified wood.
Union  Agate.
Wallowa  Agate.
Wasco  Agate, amethyst, bloodstone, chalcedony, geode, jade, jasper, opal, quartz, sagenite.
Wheeler  Agate.

### Pennsylvania:
Adams  Cuprite.
Bedford  Calcite, quartz, spar.
Berks  Calcite, epidote, feldspar, garnet, hematite, kyanite, magnetite, quartz crystal, unakite, zeolite.
Bucks  Galena, sphalerite.
Carbon  Autunite, carnotite.
Chester  Anglesite, azurite, feldspar, garnet, goethite, hornblend, kyanite, magnetite, malachite, martite, phlogopite, pyrite, pyromorphite, pyrrhotite, quartz, quartz crystal, sphalerite, stibnite, sulfur, wulfenite.
Cumberland  Phosphorite.
Delaware  Actinolite, apatite crystal, deweylite, feldspar, garnet.
Lancaster  Actinolite, aragonite, calcite, chalcedony, chromite, dolomite, hematite, malachite, marcasite, quartz crystal, pyrite, serpentine.
Lebanon  Anthophyllite, apophylite, biotite, garnet, hematite, magnetite, serpentine.
Lehigh  Calamine, corundum, goethite, greenockite, jasper, quartz crystal, sphalerite.
Monroe  Quartz crystal.
Montgomery  Calcite, copper and lead minerals, galena, natrolite, quartz crystal, sphalerite, stibnite, zeolite.
Northampton  Calcite, graphite, gummite, limonite, serpentine, talc, uraninite.
Perry  Travertine.
Schuylkill  Chlorite, galena, pyrite, quartz crystal, siderite, sphalerite.
Somerset  Quartz.
Westmoreland  Limonite, pyrite.
York  Limonite, pyrite.

### Rhode Island:
Providence  Amethyst, beryl, fluorite, quartz crystal.

### South Carolina:
Anderson  Tourmaline.
Chesterfield  Topaz.
Florence  Petrified wood.

### South Dakota:
Custer  Agate, amblygonite, beryl, chalcedony, feldspar, garnet, jade, jasper, lepidolite, mica, petrified wood, quartz crystal, rose quartz, sillimanite, tourmaline.
Fall River  Agate, chalcedony, jasper, petrified wood, vanadanite.
Harding  Agate, petrified wood.
Lawrence  Garnet, Rose quartz.
Meade  Gastroliths, geode, petrified wood, selenite.
Pennington  Agate, apatite, beryl, chalcedony, feldspar, galena, garnet, jasper, lepidolite, petrified wood, rose quartz, staurolite, tourmaline.
Shannon  Agate, jasper, petrified wood.

### Tennessee:
Bedford  Agate.
Carter  Unakite.
Coffee  Agate.

### Texas:
Brewster  Agate, amethyst, carnelian, chalcedony, citrine, fire opal, jasper, labradorite, moonstone, novaculite, petrified wood, quartz.
Burnet  Garnet, topaz.
Culberson  Agate.
De Witt  Agate, jasper, petrified wood.
Duval  Agate, jasper, petrified wood.
El Paso  Agate.
Fayette  Petrified wood.
Gillespie  Amethyst, garnet, petrified wood, topaz.
Gonzales  Petrified wood.
Hidalgo  Agate.
Hudspeth  Agate.

*State; County*  *Gem Materials*

Jeff Davis  Adularia, agate, amethyst, carnelian, chalcedony, citrine, jasper, moonstone, opal, petrified wood, quartz crystal.
Lee  Petrified wood.
Live Oak  Agate, petrified wood.
Llano  Amethyst, garnet, quartz crystal, topaz, tourmaline.
McMullen  Petrified wood.
Mason  Amazonite, cassiterite, fluorite, quartz, topaz.
Pecos  Agate.
Potter  Petrified wood.
Presidio  Agate, amethyst, carnelian, chalcedony, citrine, jasper, moonstone, opal, petrified wood, quartz crystal.
Reeves  Agate, amethyst, carnelian, chalcedony, citrine, jasper, moonstone, opal, petrified wood, quartz crystal.
Taylor  Smoky quartz, topaz.
Terrell  Agate.
Travis  Topaz.
Trinity  Petrified wood.
Walker  Petrified wood.
Webb  Agate, jasper.
Zapata  Agate, jasper.

## Utah:
Beaver  Agate, obsidian, quartz crystal.
Box Elder  Variscite.
Carbon  Agate.
Emery  Agate, chalcedony, obsidian, petrified wood.
Garfield  Agate, barite, dinosaur bone, jasper, onyx, petrified wood.
Grand  Agate, dinosaur bone, jasper, petrified wood.
Iron  Agate, geode.
Juab  Agate, geode, jasper, topaz.
Kane  Agate, petrified wood.
Millard  Fossils, jasper, obsidian.
Salt Lake  Agate, onyx.
Sevier  Agate.
Tooele  Agate, geode, obsidian.
Utah  Onyx, variscite.
Washington  Agate, alabaster, azurite, jasper.
Wayne  Agate, barite, dinosaur bone, jasper, petrified wood.

## Vermont:
Windsor  Actinolite, magnetite, pyrite, talc.

*State; County*  *Gem Materials*

## Virginia:
Amelia  Albite, amazonite, cleavelandite, garnet.
Madison  Blue quartz, epidote, unakite.
Page  Epidote, jasper, onyx.
Prince Edward  Amazonite, amethyst, kyanite.
Rockbridge  Unakite.
York  Shark teeth, whalebone.

## Washington:
Benton  Petrified wood.
Chelan  Thulite.
Cowlitz  Carnelian, sardonyx.
Douglas  Jadeite, thulite.
Kittitas  Agate, chalcedony, jasper, petrified wood.
Klickitat  Agate, jasper, petrified wood.
Lewis  Carnelian, sardonyx.
Snohomish  Petrified wood.
Yakima  Carnelian, sardonyx.

## West Virginia:
Hardy  Aragonite, stilbite.
Mineral  Fossils.

## Wisconsin:
Ashland  Agate, jasper.
Bayfield  Agate.
Clark  Agate, jasper.
Douglas  Agate.
Iron  Agate.

## Wyoming:
Albany  Agate, petrified wood.
Big Horn  Agate.
Carbon  Agate, jade, jasper, petrified wood.
Fremont  Actinolite, agate, aventurine, garnet, jade, jasper, petrified wood, rhodonite, sapphire, serpentine.
Goshen  Jasper.
Johnson  Petrified wood.
Lincoln  Agate, petrified wood.
Natrona  Agate, amazonite, jade, petrified wood.
Platte  Agate.
Sweetwater  Agate, chalcedony, corundum, eden wood, jade, jasper, moss agate, petrified wood.
Uinta  Petrified wood.

# INDEX

American Antiquity Act, 61-63
American Federation of Mineralogical Societies, 142
American Numismatic Society, 81, 92
*Argosy,* 162
Arkansas, buried treasure in, 3

Bahamas Ministry of Tourism, 160
Barber, Charles E., 82
Battery, check meter system, 16, 21, 22, 25, 26
Battlefields, 102, 103
Beaches, 2, 8, 31, 39, 64, 67-70, 72, 93
Blanchard, Chet, 56
Bludworth Marine, 157, 158
Bottles
 collecting of, 93-99
 cleaning of, 99
 digging of, 39, 61, 96, 99
 hunting tools and, 99
 literature on, 100
 sites of, 95, 96, 99
Bowling Green Park, New York City, 3
Buildings, abandoned, 2, 5, 46, 54, 57-60
Bullets, 4, 102
Bureau of Land Management, 62, 117-119, 134, 135
Bureau of Mines and Mineral Resources, 143
*The Button Sampler,* 104
Buttons, 95, 104

California Division of Mines and Geology, 131
Cape Cod National Seashore Project, 70
Cape Hatteras, 148
Cape Henlopen, 70
Carnivals, sites of, 71
Cartridges, 102
Carver, George Washington, 86
Coinage Act of 1965, 87
Coins, 1, 4, 17, 26, 28, 37, 46, 64-92
 cleaning of, 75, 76
 for collectors, 2, 76, 85-87, 89, 90, 161
 commemorative, 85, 86
 condition of, 66, 68, 69, 79, 80
 copper, 69, 78, 80
 corrosion of, 66, 69, 75
 depth under soil, 65, 67, 72, 74, 78
 dimes, 66, 68, 76, 81, 82, 87
 foreign, 4, 70, 146
 gold coins, 66, 69, 70, 76, 89
 grades of, 76-80, 84, 85
 half dollars, 66, 82, 84
 hot spots for, 64, 68-72, 74, 84, 90
 mint marks on, 76-82, 84
 nickels, 66, 76, 79, 81, 87
 pennies, 66, 68, 78-82, 87
 publications on, 91, 92
 quarters, 66, 81, 82, 87
 recovering of, 73-75
 selling of, 80, 85
 silver, 3, 66, 68-70, 76, 80, 82, 84-87
 storing of, 75, 76
 tokens, 71, 89-91
 value of, 67, 76-82, 87, 88, 161
Coinshooting, 4, 5, 38, 43-46, 61, 64-69, 72, 74, 78, 82, 103
Collectibles, 93-109
Compass Electronics Corp., 19, 22, 23, 121, 123
*Complete Book of American Country Antiques,* 102

Deeds, buried, 4
Delaware Bay, 70
Detectors, metal
 audio response, 11, 16, 21
 balance, 8, 21, 26
 batteries, 6, 16, 17, 21, 25, 26, 44
 beat frequency, 4, 9, 10, 14, 26
 components, 6
 control box, 2, 6, 8, 10, 13, 14, 16, 18, 21, 26, 44
 control knobs, 6, 14, 21, 22, 25, 36
 depth penetration, 2, 7, 8, 12, 21, 22, 26, 27, 29
 "discriminator," 18, 19, 26, 27, 57
 drift in, 28, 29, 122
 earphones, 17, 21, 22, 25, 26, 41, 42
 electronic circuitry, 6, 25, 36
 indicator light, 19
 metrotech, 26
 models, 6-12, 21-26, 28, 29, 34, 36

operation of, 2, 6, 13, 35-38, 42, 43
prices of, 2, 4, 5, 8, 13, 21, 22, 25, 26, 64, 122
range control, 36
search coil, 9, 11-14, 18, 19, 26, 29, 36, 37, 122
search head, 6, 8, 10, 12, 18, 21, 22, 25
search loop, 2, 5, 8, 13, 14, 16, 18, 21, 22, 26
second hand, 20
shaft, 8, 17, 18, 22
signal, 4-6, 10, 11, 16, 18, 22, 26, 28, 42
speed handle, 18, 19
stability, 28, 42
storage of, 43, 44
swivel head, 57
transmitter-receiver, 2, 6, 9, 10, 12-14, 16, 21
tuning, 35, 36, 37
underwater models, waterproof, 2, 14, 18, 21, 22, 107, 122, 154-159
weight of, 6, 8, 21, 22, 25, 26
Detectron, 23, 25
Deviney, Calvin, 153, 154
Directory of Historical Societies and Agencies in United States and Canada, 48
Divining rod, 32-34
Dumais Electronics, 38, 68
Dump searching, 12, 55, 57, 58, 94, 99

Equipment
  digging, 1
  treasure-hunting; See Tools
Excelsior Electronics Company, 25

Federal Reserve Notes, 86
Fisher, Dr. Gerhardt, 6, 9
Fishers Manufacturing Company, 26, 154-156
Fishmaster Products, 159
Flea markets, 93
Florida, coastal waters of, 144, 146, 147
Florida Board of Archives and History, 145, 163
Florida coast, Spanish galleon wrecks on, 5, 69, 145, 146, 148, 149, 162
Florida Historical Society, 48
Florida Nautical Charts, 48
Florida State News Bureau, 147
Forts, 48
Foundations, of old buildings, 54, 59

Gardiner Electronics Company, 18, 26
Garrett, Charles, 74
Garrett Electronics, 9, 13, 15, 16, 18, 19, 21, 22, 122
Gem minerals, 137, 138, 139, 140, 142
  diamonds, 137
  emeralds, 140
  literature, 143
  rubies, 139, 140
  sapphires, 139, 140
  sites of, 137, 140, 143
Ghost towns, 5, 54-56, 63, 96
Glass Container Manufacturers Institute, 97
Glass insulators, 100
Glassmakers, 95, 96, 100

Gold, 110-143
  abandoned mines, 132, 133
  black magnetic sand, 2, 120, 122-124, 126, 129
  bullion, 111
  deposits, 5, 112, 119, 120, 123, 126, 130
  dry gold, 128-132
  dry washer, 130, 131
  maps, 116
  market price of, 110, 111
  mining, 129, 130
  nuggets, 2, 71, 122
  panning of, 110, 112-114, 119, 120, 123-127
  placer gold, 119, 120, 126, 127
  prospecting of, 2, 4, 5, 110-114, 116, 118-121, 124, 129, 136, 137
  sites of, 112-116, 118-120, 128, 129
  sluice box, 123-128
  Spanish, 145, 146
  stacking claims for, 133-135
  surface dredges, 127, 128
Goldak Corporation, 26, 27, 71

*Handbook of United States Coins*, 77
Heath Company, 29
Heathkit, 30
Historical
  artifacts, 4, 54, 55, 63
  relics, 2, 45, 46
  societies, 3, 48
Home-built metal detector, 29-32
  batteries, 31
  construction methods, 29-31
  deep type, 30
  hardware, 31
  kits, 29, 31
  magnetometer, 31, 32
  price of, 29, 31
  search loop, 30
  tools, 31
  transmitter receiver system, 29, 31, 32
  unit control system, 31
Hot spots
  old foundations, 54
  ski resorts in summer, 4, 5, 71
*How-to-Test Detector Field Guide*, 10, 15, 29

Indiana Historical Society, 48
International Monetary Fund, 110
*Introduction to Treasure Hunting*, 31
Iron
  artifacts, 100-102
  workers, 101

Jamestown, Virginia, 95

Keene Engineering, 123, 125, 128, 129, 137
Kovacs Company, 158

Lagal, Roy, 10, 15, 29
Libraries, 46-50, 54
Litchfield, Connecticut, 3
Looting and vandalism, 55

# Index

Lost and found service, 107
Lots, demolition, 4
Lyons, Dr. Eugene, 148

Magnetometer, 31, 32, 153, 154
Maps, 48, 51-54, 116-118, 150
Mason, John, 98
Massachusetts, relic hunt in, 3
Maxwell Creek, California, 4
Metal locators. *See* metal detectors
Michigan Department of Natural Resources, 148
Military
    buttons, 4, 102, 104
    medals, 2
    relics, 102, 103, 152
Mineral detection, 16, 25
Mining
    permits, 110
    towns, 54, 56
Monson, Massachusetts, 54
Montana Department of Highways, 111, 125
Moulton, William, 78
Myty-Myte, 24, 25

National Association of Skin Diving Schools, 144
National Association of Underwater Instructors, 144
National Forests lands, 62, 116, 117, 119
National Forest Service, 119
National Ocean Survey, 69, 150, 151
National Resource lands, 116, 117
National Treasure Hunters League, 106
New York Commodity Exchange, 80, 87
*The New York Times,* 3, 49, 80, 89, 110

Oregon State Highway Department, 133

Pacific Northwest Instruments Inc., 18, 22
Parkersburg, West Virginia, 4
Parking lots, 71
Parks, public, 2, 8, 43, 44, 67, 68, 70, 72, 93
Pipes, clay, 4
Pitkin Glassworks, Connecticut, 96
Polo, Marco, 32
*Popular Electronics,* 30
Portrait flasks, 96
Posthole banks, 58, 60
Private property
    old homes, 70
    permit to search, 60
    searching for treasure, 60
Professional Association of Diving Instructors, 144
Prospectors Club International, 106
Public records, 48

Radio direction finder, 6
Radio Shack, 29
Rakes, Charles D., 30
Ram Publishing Company, 10, 15, 29
Revolutionary War, 95

Rings, locating of, 2, 4, 103, 104
Rockhounding, 143

Salvaging, 145, 146, 160
Scrap metal, 37
Sculpture, 3
Search Electronics, 67
Shipwrecks, 69, 70, 144-151, 154, 158
    charts of, 150
    literature on, 151
    researching sites of, 148, 151, 152
    souvenirs from, 152, 160, 162
Silver certificates, 86
Silverware, 2, 4
*Skin Diver,* 157
Sony, 32
Stiegel, Henry William, 95
Strait, Newton A., 102

Tools, 1, 38, 67, 68, 73-75, 99, 106, 118, 120, 123, 125, 126, 128, 137, 141
*Treasure Divers Guide,* 149, 159
Treasure diving, laws of, 161, 162
Treasure finds, 49, 60, 93-109
Treasure hunt, organized, 104-106
*Treasure Hunters Yearbook,* 62, 72
*Treasure Hunting Guide,* 74
Treasure hunting
    agreement form, 60
    income taxes and, 108-109
    literature on, 46, 48-50
    local laws and, 61, 62
    state-owned land and, 61
Treasure Salvors, Inc., 146, 148
Treasure sites, 23, 48, 55, 56, 60, 61, 67-71, 90, 93, 106, 145
Treasures, buried, 2, 3
Treasury Department, 86, 87
*True Treasure and Treasure World,* 106

Underwater prospecting, 107, 128, 144-163
    charts, 149, 150, 151
    divers, 149, 152, 153
    equipment used in, 153, 154, 155, 156, 157, 158, 159
    research wrecks, 148, 149, 151, 152
    salvaging, 159, 160, 161
    sites for, 144, 148, 149, 153
    sonic depth finder, 158, 159
United States Geological Survey, 53, 54, 111, 113, 114, 117, 136
United States Library of Congress, 47, 151
    National Union catalog, 47
United States Tariff Act of 1930, 93

View meter, 16, 21
Virginia City, Nevada, 54

*Wall Street Journal,* 80
White Electronics, 7, 20-22, 24, 26, 36, 121, 122
Wistar, Caspar, 95
Wright, Dr. Richard, 148